SFPE Engineering Guide to Performance-Based Fire Protection Analysis and Design of Buildings

Society of Fire Protection Engineers

 National Fire Protection Association
NFPA Quincy, Massachusetts

Society of Fire Protection Engineers
SFPE Bethesda, Maryland

Product Manager: Jim Linville
Editorial Production Services: Omegatype Typography, Inc.
Interior Design: Carol Somberg, Omegatype Typography, Inc.
Cover Design: McCusker Communications
Manufacturing Buyer: Ellen Glisker
Printer: Courier/Westford

Published by the National Fire Protection Association
One Batterymarch Park
Quincy, MA 02269-9101
Copyright © 2000
Society of Fire Protection Engineers and National Fire Protection Association

Notice Concerning Liability: While every effort has been made to achieve a work of high quality, neither SFPE, NFPA, nor the authors and contributors to this work guarantee the accuracy or completeness of or assume any liability in connection with the information and opinions contained in this work. SFPE, NFPA, and the authors and contributors shall in no event be liable for any personal injury, property, or other damages of any nature whatsoever, whether special, indirect, consequential, or compensatory, directly or indi-rectly resulting from the publication, use of or reliance upon this work.

This work is published with the understanding that SFPE, NFPA, and the authors and contributors to this work are supplying information and opinion but are not attempting to render engineering or other professional services. If such services are required, the assis-tance of an appropriate professional should be sought.

NFPA No.: SFPE-00
ISBN: 0-87765-422-0
Library of Congress Card Catalog No.: 99-069306
First Edition: 2000

Printed in the United States of America
10   9   8   7   6   5   4   3   2   1

The SFPE Task Group on Performance-Based Analysis and Design

CHAIRMAN
Eric R. Rosenbaum, P. E.
Hughes Associates, Inc.

VICE-CHAIRMAN
Michael O'Hara, P. E.
The MountainStar Group, Inc.

STEERING COMMITTEE

Douglas Beller, P. E.
National Fire
Protection Association

Allan Coutts, P. E., Ph.D.
Westinghouse Safety
Management Solutions

Richard L. P. Custer, FSFPE
Custer Powell Inc.

Daniel Gemeny, P. E.
Rolf Jensen & Associates Inc.

Christopher Marrion, P. E.
Arup Fire

Brian M. McGraw, P. E.
Hughes Associates, Inc.

**Frederick Mowrer, P. E.,
Ph.D., FSFPE**
University of Maryland

Daniel O'Connor, P. E.
Schirmer Engineering Corp.

Andrew Valente, P. E.
Rolf Jensen & Associates Inc.

MEMBERS

John F. Bender, P. E.
Office of the Maryland State
Fire Marshal

Frederick Bradley, P. E.
Gage-Babcock &
Associates Inc.

William Burrus, P. E.
City of San Antonio Building
Inspections Department

Craig Beyler, Ph.D., FSFPE
Hughes Associates, Inc.

Eugene A. Cable, P. E.
U.S. Department of
Veterans Affairs

Carol A. Caldwell, P. E.
Caldwell Consulting, Ltd.

Eric H. Cote, P. E.
Rolf Jensen & Associates Inc.

John Curry
U.S. Army Corps of Engineers

John Devlin, P. E.
Schirmer Engineering Corp.

Joshua M. Fleischer
Duke Engineering & Services

Cynthia Gier, P. E.
Henderson Engineers, Inc.

**George Hadjisophocleous,
P. Eng., Ph.D.**
National Research Council

Wayne D. Holmes, P. E.
HSB Professional Loss Control

Howard Hopper, P. E.
Underwriters Laboratories

Roland J. Huggins, P. E.
American Fire Sprinkler
Association

John E. Kampmeyer, P. E.
Triad Fire Protection, Inc.

Brian Lattimer, Ph.D.
Hughes Associates, Inc.

Ian C. MacDonald
RJ Bartlett Engineering

Rodney A. McPhee
Canadian Wood Council

Anthony J. Militello
U.S. Navy

Bijan Najafi, P. E.
SAIC

Kathy A. Notarianni, P. E.
National Institute of Standards
and Technology

Igor Oleszkiewicz, P. Eng.
National Research Council

Alan C. Parnell
Fire Check Consultants

**J. Kenneth Richardson,
P. Eng., FSFPE**
Ken Richardson Fire
Technologies Inc.

Jeffrey Shapiro, P. E.
International Code Consultants

Amal Tamim
W. R. Grace & Company

Robert J. Thompson, P. E.
U.S. Department of
Veterans Affairs

John Watts, Ph.D.
Fire Safety Institute

Christopher B. Wood, J.D.
Custer Powell Inc.

STAFF
Morgan J. Hurley, P. E.
Society of Fire Protection Engineers

Brian J. Meacham, P. E.
Society of Fire Protection Engineers

Contents

1

Introduction

1.1 Purpose

1.1.1 *The SFPE Engineering Guide to Performance-Based Fire Protection Analysis and Design of Buildings* outlines a process for using a performance-based approach in the design and assessment of building fire safety within both prescriptive and performance-based code systems.

1.1.2 The intent of the guide is threefold.

1.1.2.1 This guide provides information that both qualified engineers and authorities having jurisdiction can use to determine and document the achievement of fire safety goals for a particular project over the life of a building.

1.1.2.2 This guide identifies parameters that should be considered in performance-based analysis or design.

1.1.2.3 This guide provides a means by which engineers can develop fire protection measures that provide levels of safety deemed acceptable by the stakeholders without imposing unnecessary constraints on other aspects of building design and operation.

1.2 Fundamentals

1.2.1 This guide is written for developing and evaluating analyses by engineers practicing fire protection engineering who have a fundamental understanding of fire dynamics and who are competent in the application of scientific and engineering principles to the evaluation and design of systems and methods to protect people and their environments from the unwanted consequences of fires. Such engineers are qualified by education, training, and experience, and

- Possesses a working knowledge of the nature and characteristics of fire and related hazards, as well as a knowledge of how fires originate, develop, and spread
- Understands hazards and risks
- Understands fire prevention, manual and automatic detection, and control and suppression systems and practices
- Understands the impact of fire and fire effluents on buildings, processes, systems, and people

Unless otherwise noted, the term *engineer,* when referenced in this guide, refers to the engineer practicing fire protection engineering who is responsible for the performance-based analysis and design.

1.2.2 Many factors and fire safety parameters that are addressed in this guide might need to be considered for any given performance-based design. The factors and parameters that need to be evaluated or addressed for any given design or fire scenario will depend on the context of the analysis and the reasons for undertaking the design. Although not all factors and parameters within this guide are key considerations in every design, this guide intends to present factors and parameters as potential elements for consideration or evaluation.

1.2.3 If the process and methodology of this guide are used as intended, a written engineering document will result. This guide delineates the information gathered and employed during each step of the process as a means of establishing conclusions and recommendations.

1.2.4 The role of the engineer on a project might include the following responsibilities:

- Ascertaining stakeholder goals and objectives
- Identifying the fire hazards and risks and informing the stakeholders
- Identifying the regulatory compliance issues
- Establishing and executing the performance-based analysis
- Making fire safety recommendations consistent with stakeholder objectives and regulatory requirements

- Securing the stakeholders'—including authorities having jurisdiction—acceptance of the proposed performance-based design
- Verifying the implementation of the performance-based design via construction document development and field observations
- Developing an operations and maintenance (O&M) manual, which outlines the design conditions that must be satisfied to ensure continued acceptability of the design over the life of the building

1.2.5 The engineer and other stakeholders must understand the engineer's role on the project and how the engineer fits into the project delivery process and design team structure.

1.2.6 For effective implementation, the performance-based process should start in the concept development, feasibility, or planning stages of a project. However, this methodology can be used any time during the design and construction of a building.

1.3 Scope

1.3.1 This guide focuses on the application of scientific and engineering principles to the protection of people and property from the unwanted effects of fire. It provides a performance-based fire protection engineering approach to building fire safety analysis and design, and it provides a way to assess the effectiveness of the total building fire protection system in achieving specified fire and life safety objectives.

1.3.2 This guide presents a process for performance-based fire protection engineering for buildings. However, this design guide might be used for other applications when justified by an engineer.

1.3.3 This guide defines a performance-based process and provides references to available sources of fire protection engineering analysis, design tools, and methods; fire test methods and data; and performance criteria. However, it does not provide specific fire protection engineering analysis, design tools, methods, or data; nor does this guide provide specific performance, acceptance, or design criteria for use in the analysis and design process. (Some specific tools, methods, and criteria are provided as examples. The information cited in the examples does not necessarily constitute the correct or only information pertinent to a specific design project. See 1.5.6.)

1.3.4 This guide outlines procedures and a methodology for building fire safety designs and is intended for use by engineers. It does not detail all the engineering technology required for building fire safety design.

1.4   Use and Application

1.4.1   Performance-based design of fire protection systems may offer a number of advantages over prescriptive-based design.

(a) Performance-based design specifically addresses a building's unique aspects or uses, specific stakeholder needs, and the broader community's needs where appropriate.

(b) Performance-based design provides a basis for the development and selection of alternative fire protection options based on the project's needs (e.g., when the code-prescribed solution does not meet the stakeholders' needs).

(c) Performance-based design allows the safety levels provided by alternative design options to be compared. The comparison of options provides a mechanism for determining what level of safety, at what cost, is acceptable.

(d) Performance-based design requires the use of a variety of tools in the analysis, bringing increased engineering rigor and resulting in innovative design options.

(e) Performance-based design results in a fire protection strategy in which fire protection systems are integrated, rather than designed in isolation.

A comprehensive, performance-based engineering approach may provide more effective fire protection that addresses a specific need, in addition to improved knowledge of the loss potential.

1.4.2   Tools of performance-based fire protection engineering might include deterministic analysis techniques, probabilistic analysis techniques, application of the theory of fire dynamics, application of deterministic and probabilistic fire effects modeling, and application of human behavior and toxic effects modeling.

1.5   Technical References and Resources

1.5.1   An important part of a performance-based design is the selection and application of engineering standards, calculation methods, and other forms of scientific information that are appropriate for the particular application and methodology used. The engineer and other technically qualified stakeholders should determine the acceptability of the sources and methodologies for the particular applications in which they are used.

1.5.2   The sources, methodologies, and data used in performance-based designs should be based on technical references that are widely accepted and utilized by the appropriate professions and professional groups. Acceptance of references is often based on documents that are developed, reviewed, and validated by one of the following processes:

(a) Codes that are developed through an open consensus process conducted by recognized professional societies, code-making organizations, or governmental bodies

(b) Technical references that are subject to a peer review process and published in widely recognized journals, conference reports, or other publications

(c) Resource publications, such as *The SFPE Handbook of Fire Protection Engineering*,[1] that are widely recognized technical sources of information

1.5.3   The following factors are helpful for determining the acceptability of the individual method or source:

(a) Extent of general acceptance in the relevant professional community. Indications of this acceptance include peer-reviewed publication, widespread citation in the technical literature, and adoption by or within a consensus document.

(b) Extent of documentation of the method, including the analytical method itself, assumptions, scope, limitations, data sources, and data reduction methods.

(c) Extent of validation and analysis of uncertainties. This process includes a comparison of the overall method with experimental data to estimate error rates as well as an analysis of the uncertainties of the input data, the uncertainties and limitations in the analytical method, and the uncertainties in the associated performance criteria.

(d) Extent to which the method is based on sound scientific principles.

(e) Extent to which the proposed application is within the stated scope and limitations of the supporting information, including the range of applicability for which there is documented validation. Factors such as spatial dimensions, occupant characteristics, and ambient conditions might limit valid applications.

1.5.4   In many cases, a method will be developed from and include numerous component analyses. These component analyses should be evaluated using the same factors applied to the overall method as outlined in 1.5.3.

1.5.5   A method to address a specific fire safety issue within documented limitations or validation regimes might not exist. In such a case, sources and calculation methods can be used outside of their limitations, provided the engineer recognizes the limitations and addresses the resulting implications.

1.5.6   The technical references and methodologies used in a performance-based design should be closely evaluated by the engineer, technically qualified stakeholders, and possibly a third party reviewer. The strength of the technical justification should be judged using criteria listed in Section 1.5.3. This justification might be strengthened by the presence of data obtained from fire testing.

1.5.7   Throughout the text, this guide refers to selected technical references and resources used by the fire protection community at the time of publication. This list of documents is not all inclusive nor is it an endorsement of specific sources or methodologies that might be contained within the documents. It is still the responsibility of the engineer and stakeholders to determine the acceptability of these technical reference sources or methodologies in a particular performance-based design.

Reference Cited

1. Dinenno, P., Ed. *The SFPE Handbook of Fire Protection Engineering,* 2nd Ed., National Fire Protection Association, Quincy, MA, 1995.

2

Glossary

This chapter defines terms as they will be used in this document.

Authority Having Jurisdiction (AHJ) An organization, office, or individual responsible for approving designs, equipment, installations, or procedures.

Building Any structure used or intended for supporting or sheltering any use or occupancy.

Building Characteristics A set of data that provides a detailed description of a building, such as building layout (geometry), access and egress, construction, building materials, contents, building services, and fire safety (hardware) systems (see 8.2.3.3).

Client The party for which professional services are rendered.

Confidence Interval A statistical range with a specified probability that a given parameter lies within the range.

Design Criteria See Performance Criteria.

Design Fire Curve An engineering description of the development of a fire for use in a design fire scenario. Design fire curves might be described in terms of heat release rate versus time, or in other terms (see 8.5.4).

Design Fire Scenario A set of conditions that defines or describes the critical factors for determining outcomes of trial designs (see 8.5).

Design Objective   A description of the performance benchmark against which the predicted performance of a design is evaluated.

Deterministic Analysis   A methodology based on physical relationships derived from scientific theories and empirical results that for a given set of initial conditions will always produce the same result or prediction. In a deterministic analysis, a single set of input data will determine a specific set of output predictions.

Final Design   The design that is selected from among the successful trial designs and chosen for implementation.

Fire Characteristics   A set of data that provides a description of a fire (see 8.2.3.5).

Fire Model   A physical or mathematical procedure that incorporates engineering and scientific principles in the analysis of fire and fire effects to simulate or predict fire characteristics and conditions of the fire environment.

Fire Protection Engineering Design Brief   A document summarizing agreed upon performance criteria and methods that will be used to evaluate trial designs (see Chapter 11).

Fire Safety Goals   Desired overall fire safety outcome expressed in qualitative terms.

Fire Scenario   A set of conditions that defines the development of fire and the spread of combustion products throughout a building or part of a building (see 8.2).

Frequency   The number of times an event occurs within a specified time interval.

Hazard   A possible source of danger that can initiate or cause undesirable consequences if uncontrolled.

Objective   A requirement of the fire, building, system, or occupants that needs to be fulfilled in order to achieve a fire safety goal. Objectives are stated in more specific terms than goals. In general, objectives define a series of actions that make the achievement of a goal more likely.

Occupant Characteristics   A set of data that describes conditions, abilities, or behaviors of people before and during a fire (see 8.2.3.4).

Performance-Based Code   A code or standard that specifically states its fire safety goals and references acceptable methods that can be used to demonstrate compliance with its requirements. The document might be phrased as a method for quantifying equivalencies to an existing prescriptive-based document, might identify one or more prescriptive documents as approved solutions, or might specify performance criteria without referencing prescriptive requirements. The document allows the use of any solution that demonstrates compliance.

Performance-Based Design   An engineering approach to fire protection design based on (1) established fire safety goals and objectives; (2) deterministic and probabilistic analysis of fire scenarios; and (3) quantitative assessment of design alternatives against the fire safety goals and objectives using accepted engineering tools, methodologies, and performance criteria.

Performance-Based Design Option   An option within a code where compliance is achieved by demonstrating through an engineering analysis that a proposed design will meet specified fire safety goals. More specifically, fire safety goals and objectives are translated into performance objectives and performance criteria. Fire models, calculations, and other verification methods are used in combination with the building design specifications, specified fire scenarios, and specified assumptions to determine whether the performance criteria have been met, which proves compliance with the code under the performance-based design option.

Performance Criteria   Criteria stated in engineering terms with which the adequacy of any developed trial designs will be judged.

Prescriptive-Based Code   A code or standard that prescribes fire safety for a generic use or application. Fire safety is achieved by specifying certain construction characteristics, limiting dimensions, or protection systems without referring to how these requirements achieve a desired fire safety goal.

Prescriptive-Based Design Option   An option within a code where compliance is achieved by demonstrating compliance with specified construction characteristics, limits on dimensions, protection systems, or other features.

Probabilistic Analysis   An evaluation of the fire losses and fire consequences, which includes consideration of the likelihood of different fire scenarios and the inputs that define those fire scenarios.

Probability   The likelihood that a given event will occur. Statistically, the number of actual occurrences of a specific event divided by the total number of possible occurrences. Probabilities are inherently unitless and expressed as a number between zero and one, inclusive.

Project Scope   An identification of the range or extent of the design matter being addressed, including any specific limits of a performance-based design. The project might be a subset of a larger development, evaluation, or design effort (e.g., one part of the building design process), or it might be a stand-alone fire safety analysis and design project.

Risk   In the classical engineering sense, the product of the potential consequences and the expected frequency of occurrence. Consequences might include occupant death, monetary loss, business interruption, or environmental damage. The frequency of occurrence could be an estimate of how often the projected loss might occur.

Safety Factor   Adjustments made to compensate for uncertainty in the methods, calculations, and assumptions employed in the development of engineering designs.

Stakeholder   One who has a share or an interest in an enterprise. Specifically, an individual (or a representative) interested in the successful completion of a project. Reasons for having an interest in the successful completion of a project might be financial or safety related (see 4.2.4).

Stakeholder Objective   A statement of a stakeholder's level of acceptable or sustainable loss.

Trial Design   A fire protection system design that is intended to achieve the stated fire safety goals and that is expressed in terms that make the assessment of these achievements possible.

Uncertainty   The amount by which an observed or calculated value might differ from the true value.

Verification   Confirmation that a proposed solution meets the established fire safety goals. Verification of a performance-based design might involve alternate computer fire models to reproduce results or full-scale fire testing.

Worst Case Scenario   A scenario resulting in the worst consequence as defined by the stakeholders or a code. The criteria must be explicitly stated because worst case conditions for life safety and property protection might be incompatible. (Also see Worst Credible Fire.)

Worst Credible Fire   For a specific site, a fire, as defined by the stakeholders or a code, that can be reasonably expected to result in unfavorable consequences equal to or less severe than those resulting from a worst case scenario.

3

Overview of the Performance-Based Fire Protection Analysis and Design Process

3.1 General

3.1.1 This guide, which is applicable to new and existing buildings, can be used for the design of fire protection measures to achieve stated fire and life safety objectives, to support the development of alternatives to prescriptive-based code requirements, or to evaluate the building fire safety design as a whole.

3.1.2 Performance-based fire protection analysis and design is one element in the process of building design, construction, and operation.

3.1.2.1 The performance-based design process most appropriately begins during the feasibility or conceptual design phase when key decisions are being made (see Section 1.4). The earlier the engineer becomes involved in the building design and construction process, the greater the benefits that can be realized. These benefits include the following:

- Design flexibility
- Innovation in design, construction, and materials
- Equal or better fire safety
- Maximization of the benefit/cost ratio

A representative basic building design and construction process is outlined in Figure 3–1.

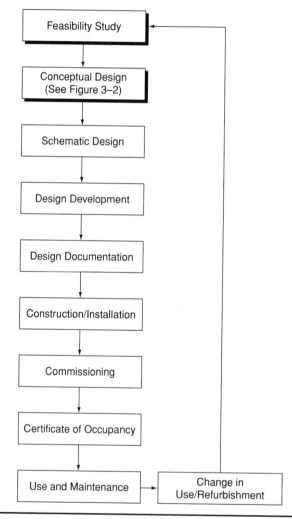

FIGURE 3–1   *Basic Building Design and Construction Process. Note: Feasibility Study and Conceptual Design are highlighted due to the importance of involving the engineer early in the overall design process (see 3.1.2.1).*

3.1.3 During the design development and construction documentation phases, the conceptual designs evolve into detailed system designs and the plans and specifications that will be used for bidding on and constructing the building. The design development and construction documentation phases are not described in this guide. However, the coordination of design documentation is described in Chapter 12. Coordination between disciplines is critical during these phases in order to ensure proper interaction between systems.

3.1.4 The concepts utilized in the performance-based design process, including the limitations and assumptions, must be reviewed by the stakeholders throughout the design process. Design changes resulting from value engineering or other design procedures must be incorporated into the performance-based design analysis. For example, changes to systems that seem unrelated to fire safety, such as the bathroom exhaust systems or operable windows, could have an effect on a performance-based design incorporating smoke control systems.

3.1.5 The commissioning of fire protection systems and the review of their installation, which validates that the installed fire protection systems meet the proposed intent of the design, are essential to the level of fire safety provided in the structure. The engineer should be involved in the production and review of design documents, the review of shop drawings, field inspections, and acceptance testing of the fire protection systems.

3.1.6 After the building is commissioned and the certificate of occupancy is issued, the building's owner should assure that the building is used and maintained in accordance with the fire protection concepts incorporated in the original performance-based design. This assurance can be accomplished by the implementation of the operation and maintenance criteria specified in the O&M manuals for the building, which should be developed by the engineer (see Chapter 12). The stakeholders must recognize that a performance-based design might impose constraints on the future of the building. If a change of use or occupancy is inconsistent with the original design assumptions and conditions, the design process as illustrated in Figure 3–1 should be repeated.

Due to the unique assumptions and fire protection concepts that might be incorporated in a performance-based design, the development and maintenance of accurate documentation is critical. The building's owner, the authority having jurisdiction, the engineers, and other stakeholders involved in the operation, use, and modification of a building might change during the life of a building. Complete, accurate documentation provides a way for future stakeholders to understand and function within the limitations of the original design assumptions and conditions.

3.2   Performance-Based Design Process

3.2.1   In the performance-based design process, many acceptable approaches that will efficiently address and resolve a performance-based design issue are available. For example, the stakeholders can rank trial designs based on financial considerations, desired building features, and other factors. The engineer could then evaluate the highest ranking trial design against the design fire scenarios to determine if the trial design meets the performance criteria. When outlining a process, this guide does not intend to promote one process more than other alternatives, but it does intend to identify the process steps that should guide a performance-based design in a complete and comprehensive manner. The recommended steps in the performance-based design process are discussed in Section 3.2.2–Section 3.2.11 and Chapter 4–Chapter 12. The flowchart shown in Figure 3–2 is used in Chapter 4–Chapter 12. The subject of each chapter is highlighted in the flowchart as appropriate. The flowchart indicates where each chapter fits in the overall performance-based design process.

3.2.2   Defining Project Scope   The first step in a performance-based design is to define the scope of the project (see Chapter 4). Defining the scope consists of identifying and documenting the following:

- Constraints on the design and project schedule
- The stakeholders associated with the project
- The proposed building construction and features desired by the owner or tenant
- Occupant and building characteristics
- The intended use and occupancy of the building
- Applicable codes and regulations

An understanding of these items is needed to ensure that a performance-based design meets the stakeholders' needs.

3.2.3   Identifying Goals   Once the scope of the project is defined, the next step in the performance-based design process is the identification and documentation of the fire safety goals of various stakeholders (see Chapter 5). Fire safety goals might include levels of protection for

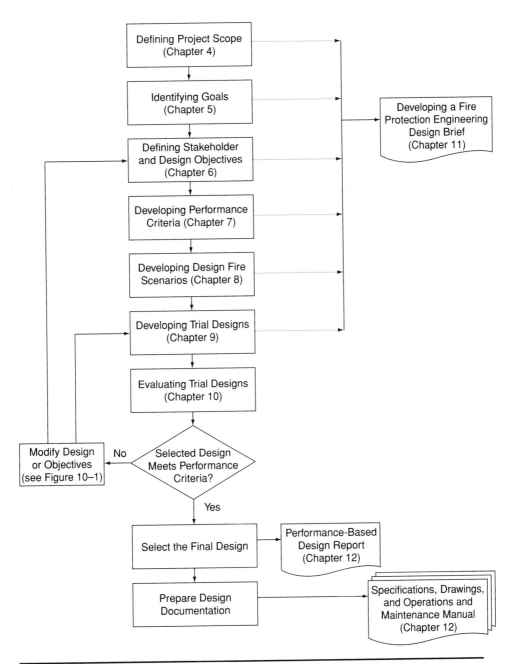

FIGURE 3–2   *Steps in the Performance-Based Analysis and the Conceptual Design Procedure for Fire Protection Design*

people and property, or they might provide for continuity of operations, historical preservation, and environmental protection. Goals might be unique, depending on the stakeholders' needs and desires, for different projects.

The stakeholders should decide which goals are the most important for the project. To avoid problems later in the design process, all stakeholders should be aware of and agree to the goals before proceeding with the performance-based design process.

3.2.4 Defining Stakeholder and Design Objectives

The third step in the design process is the development of objectives (see Chapter 6). The objectives are essentially the design goals that have been further refined into values that can be quantified in engineering terms. Objectives might include mitigating the consequences of a fire expressed in terms of dollar values, loss of life, other impacts on property operations, or maximum allowable conditions, such as the extent of fire spread, temperature, and the spread of combustion products.

3.2.5 Developing Performance Criteria

The fourth step in the design process is the development of performance criteria to be met by the design. These criteria are a further refinement of the design objectives, and they are numerical values to which the expected performance of the trial designs can be compared (see Chapter 7). Performance criteria might include threshold values for the temperatures of materials, gas temperatures, carboxyhemoglobin (COHb) levels, smoke obscuration, and thermal exposure levels.

3.2.6 Developing Design Fire Scenarios

Once the performance criteria have been established, the engineer will develop and analyze design alternatives to meet performance criteria. The first part of this process is the identification of possible fire scenarios and design fire scenarios (see Chapter 8).

Fire scenarios are descriptions of possible fire events, and they consist of fire characteristics, building characteristics, and occupant characteristics. The fire scenarios identified will subsequently be filtered (i.e., combined or eliminated) into a subset of design fire scenarios against which trial designs will be evaluated.

3.2.7 Developing Trial Designs

Once the project scope, performance criteria, and design fire scenarios are established, the engineer develops preliminary designs, referred to as trial designs, intended to meet the project requirements (see Chapter 9).

The trial design(s) includes proposed fire protection systems, construction features, and operations that are provided in order for a design to meet the performance criteria when evaluated using the design fire scenarios.

The evaluation method should also be determined at this point. The evaluation methods used must be appropriate for the situation and agreeable to the stakeholders.

3.2.8 Developing a Fire Protection Engineering Design Brief   At this point in the process, a Fire Protection Engineering Design Brief should be prepared and provided to all stakeholders for their review and concurrence (see Chapter 11). This brief should document the project scope, goals, objectives, trial designs, performance criteria, design fire scenarios, and analysis methods. Documenting and agreeing on these factors at this point in the design process will help avoid possible misunderstandings later.

3.2.9 Evaluating Trial Designs   Each trial design is then evaluated using each design fire scenario. The evaluation process is described in detail in Chapter 10. The evaluation results indicate whether the trial design will meet the performance criteria.

Only trial design(s) that meet the performance criteria can be considered as final design proposals. Yet, the performance criteria may be revised with the stakeholders' approval. The criteria cannot be arbitrarily changed to ensure that a trial design meets a criterion, but they can be changed based on additional analysis and the consideration of additional data.

3.2.10 Selecting the Final Design   Once an acceptable trial design is identified by the evaluation, it can be considered for the final project design. If multiple trial designs are evaluated, further analysis will be needed to select a final design. The selection of an acceptable trial design for the final design might be based on a variety of factors, such as financial considerations, timeliness of installation, system and material availability, ease of installation, maintenance and use, and other factors.

3.2.11 Design Documentation   Once the final design is identified, design documents need to be prepared (see Chapter 12). Proper documentation will ensure that all stakeholders understand what is necessary for the design implementation, maintenance, and continuity of the fire protection design.

The documentation should include the Fire Protection Engineering Design Brief, a performance design report, detailed specifications and drawings, and a building operations and maintenance manual.

3.3   Application and Use

3.3.1   The performance-based design process described in this guide can be used in conjunction with prescriptive-based codes, with performance-based codes, or as a stand-alone engineering analysis and design effort.

3.3.2   Use with Prescriptive-Based Codes

3.3.2.1   Prescriptive-based codes provide requirements for broad classifications of buildings. These requirements are generally stated in terms of fixed values, such as maximum travel distance, minimum fire resistance ratings, and minimum features of required systems (e.g., detection, alarm, suppression, and ventilation).

In most prescriptive-based codes, however, an *alternative methods and materials* or *equivalency* clause appears that permits, at the discretion of the authority having jurisdiction (AHJ), the use of alternative means to meet the intent of the prescribed code provisions. This provides an opportunity for a performance-based design approach via a standardized methodology. Through performance-based design, a building design can be evaluated, and its compliance with the implicit or explicit intent of the applicable code can be checked.

3.3.2.2   When employing the alternative methods and materials or equivalency clause, it is important to identify the prescriptive-based code provision being addressed (scope), to provide an interpretation of the intent of the provision (goals and objectives), to provide an alternative approach (trial design), and to provide engineering support for the suggested alternative (evaluate trial design). This guide provides a framework for addressing these issues in a logical and consistent manner.

3.3.2.3   Trial designs require a comparison of the performance of the design features required by a prescriptive-based code with the performance resulting from the trial design. Using prescribed features as a baseline for comparison (i.e., as the performance criteria), the evaluation can then demonstrate whether a trial design offers the intended level of performance. A comparison of the safety provided can be the basis for establishing equivalency of performance-based designs.

3.3.2.4 When a trial design does not meet or exceed the performance level prescribed by code, failure of the trial design could be assumed. Further review of the evaluation might reveal excesses in the prescribed code fire protection approach. These excesses might exist when the performance of a prescribed code design provides a sufficiently large margin of safety beyond that required by the performance criteria.

3.3.3 Use with Performance-Based Codes

3.3.3.1 Performance-based codes establish acceptable or tolerable levels of hazard or risk for a variety of health, safety, and public welfare issues in buildings. Unlike prescriptive-based codes, however, the levels of acceptable or tolerable risk are often expressed qualitatively, such as "safeguard people from injury due to fire" and "give people adequate time to reach a safe place," rather than being expressed as specific construction requirements.

3.3.3.2 In performance-based codes, the code writers are essentially qualifying (and sometimes quantifying) the level of risk acceptable to society (the stakeholders or users of the code). Because the definition of acceptable societal risk is more difficult than an individual stakeholder's agreement with the acceptable level of risk, several countries handle this issue by stating fire safety goals as broad social objectives, such as "safeguard people from injury due to fire," and by providing more detailed performance or functional objectives, such as "give people adequate time to reach a safe place."

3.3.3.3 The use of the terms *adequate* and *reasonable* permit design flexibility and provide general guidance, without stating specifics, on the level of risk society or a specific community is willing to accept. It is assumed that if a *reasonable* approach has been taken to determine "an adequate time to reach a safe place," the recommended safety measures will be *acceptable*.

3.3.3.4 Compliance with performance-based codes is attained through the use of either a prescriptive-based code, which has been deemed to comply as an acceptable solution, or a performance-based design approach, which provides an acceptable method for developing an acceptable solution.

3.3.4   Use as a Stand-Alone Methodology

3.3.4.1   In many cases, the basis of an analysis and design project will be prescriptive-based codes. However, additional or complementary fire safety goals and objectives might be identified, thus requiring additional fire protection engineering analysis and design.

For example, although property protection and the continuity of operations might be goals of a building's owner or insurer, they might not be fully addressed in local building and fire codes. The performance-based design process can be used to identify and address any additional goals.

3.3.4.2   To address these additional or complementary goals, the performance-based design process outlined in this guide provides a structured approach to a variety of fire safety issues as well as design flexibility.

3.4   Levels of Application

3.4.1   The performance-based design process can be used to evaluate and recommend fire protection options at the subsystem-performance level, the system-performance level, or the building-performance level. At any level, the results might be evaluated on a comparative or absolute basis.

3.4.2   A subsystem-performance evaluation typically consists of a simple comparative analysis in which it is required to demonstrate that a selected subsystem provides equivalent performance to the performance specified by a prescriptive-based code. At this level, one subsystem is evaluated in isolation.

3.4.3   A system-performance evaluation might consist of a comparative or absolute analysis. A system-performance evaluation is used when more than one fire protection system or feature is involved. It is more complex than a subsystem evaluation because the analysis must account for the interaction between various subsystems.

3.4.4   In a building-performance analysis, all subsystems used in the protection strategy and their interactions are considered. A performance-based design that analyzes total building fire safety can provide more comprehensive solutions than subsystem- or system-performance analyses because the entire building–fire–target interaction is evaluated.

3.4.5 Section 10.2 provides additional information on the levels of analysis.

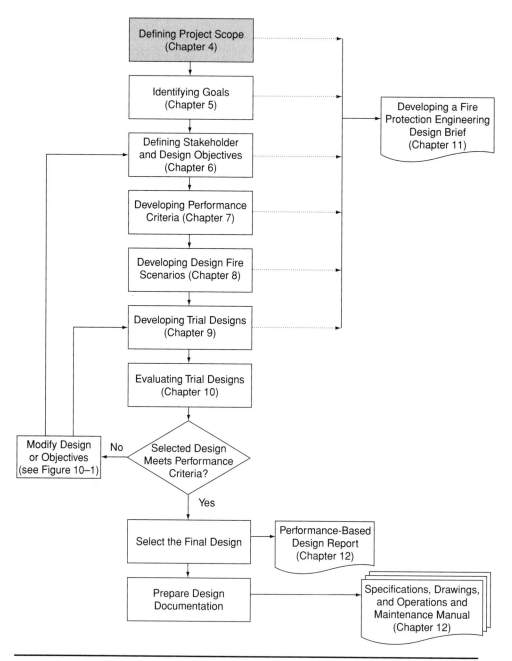

FIGURE 4–0 *Overview of the Performance-Based Design Process, Defining Project Scope*

4

Defining Project Scope

4.1   The first step in a performance-based analysis or design (see Figure 4–0) is the determination of the scope of the project.

4.2   Project Scope

4.2.1   The project scope is an identification of the boundaries of the performance-based analysis or design. The boundaries of the design might include the following: building use, design intent, project constraints, design and construction team organization (traditional versus design build), project schedules, and applicable regulations.

4.2.2   The scope might include the following elements:
- Specific fire protection system components, a partial building, a whole building, or several buildings
- New construction or renovation of an existing building either for a change in use or modernization
- Repairs to a partial building, a whole building, or several buildings

4.2.3   In addition to physical limitations on the scope, budgetary parameters can affect the cost benefit of a performance-based design.

Therefore, the available budget for the analysis and possible solutions should also be determined.

4.2.4 The scope of the project might vary depending on the perspective of the design participant. Therefore, all stakeholders in the project should be identified. The stakeholders should establish goals and objectives. The following list includes possible stakeholders:

- Building owner
- Building manager
- Design team
- Authorities having jurisdiction
  - Fire
  - Building
  - Insurance
- Accreditation agencies
- Construction team
  - Construction manager
  - General contractor
  - Subcontractors
- Tenants
- Building operations and maintenance
- Emergency responders

The stakeholders might have similar, different, or additional goals and objectives; therefore, the engineer must identify them in order to obtain their acceptance of the performance-based strategies used in the process. Concurrence of the stakeholders at the beginning improves the chances that the performance-based design alternatives will be accepted.

4.3   Submittal Schedule

4.3.1 A submittal schedule should be established as part of the project scope. It might be a stand-alone document or part of another project document.

4.3.2 The submittal schedule should list each project deliverable and who must approve the document. Approvals by multiple stakeholders might be required; however, every document does not have to be approved by all stakeholders.

4.3.3 For some projects, the approval order for each respective stakeholder might be important. If this is the case, then this information should be included on the submittal schedule.

4.3.4 The submittal schedule may specify who gets a copy of project documentation, intermediate deliverables, and when they are expected.

4.4   Issues That Should Be Considered in the Project Scope

4.4.1   The engineer must understand a building's proposed use and the characteristics of the occupants when undertaking a performance-based analysis or design. The building's functional, geometric, and operational characteristics will be the basis for developing fire scenarios.

4.4.2   In addition to building characteristics and occupant characteristics, other issues that might affect design performance should be considered. The following list is intended to be representative, but it should not be considered all-inclusive.

4.4.2.1   *Location of the Property.* The location of the property, site conditions, and the locations of adjacent properties should be determined.

4.4.2.2   *Fire Service Characteristics.* The location, the expected response time, the operating procedures, and the capabilities of the fire service and applicable emergency responders should be determined.

4.4.2.3   *Utilities.* The location and the capacity of site utilities, such as drainage, fuel, water, and electric supply, should be determined.

4.4.2.4   *Environmental Considerations.* Land use planning, effluent production, wetlands, zoning classifications, and pollution considerations should be considered.

4.4.2.5   *Historical Preservation.* This element might need to be considered if it affects subsequent objectives and design options.

4.4.2.6   *Building Management and Security.* The planned management and security schemes for the building should be ascertained.

4.4.2.7   *Economic and Social Value of the Building.* Economic value could include tax base or employment considerations. Social value could include historic, public assembly, or religious significance.

4.4.2.8   *The Project Delivery Process.* Several forms of project organization and delivery exist. Traditional architect- or engineering-lead projects, design build, fast track, and many other delivery methods will affect the development, implementation, and evaluation of the performance-based design.

4.4.2.9   *Applicable Regulations.* Identify the appropriate codes, regulations, and insurance requirements for the performance-based analysis.

4.4.2.10   When a feature is not known, a reasonable assumption may be made. However, the stakeholders might need to take steps to ensure that the assumption is valid during the life of the building. Any assumptions must be documented (see Chapter 11).

4.4.3   The above features may not be determined prior to the performance-based design. However, even if they were predetermined, they may be flexible and may be changed as part of the performance-based design. The engineer should consult with other stakeholders to determine the flexibility of the items identified as within the scope of the project.

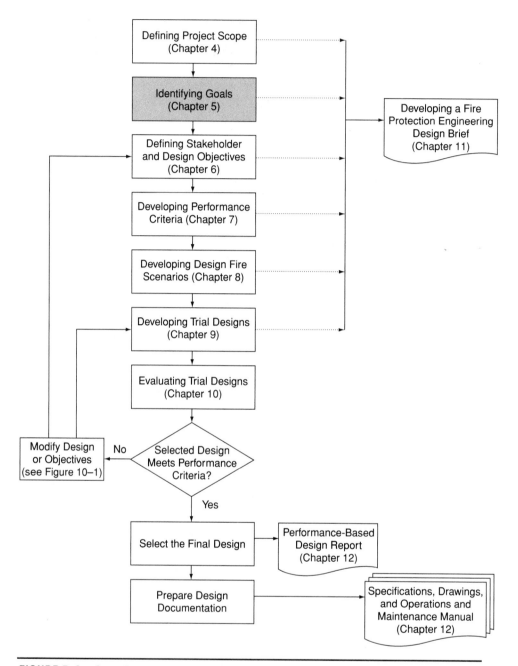

FIGURE 5–0 *Overview of the Performance-Based Design Process, Identifing Goals*

5

Identifying Goals

5.1 The next step in the performance-based analysis and design process is the determination of the fire safety goals of interest to stakeholders (see Figure 5–0) and the prioritization of them for the specific project being addressed. Goals are identified through discussions with the stakeholders and a review of background materials.

5.2 Goals for Fire Safety. The following list presents four interrelated fundamental goals for fire safety:

(a) **Provide life safety for the public, building occupants, and emergency responders.** Minimize fire-related injuries, and prevent undue loss of life.

(b) **Protect property.** Minimize damage to property and cultural resources from fire. Protect building, contents, and historical features from fire and exposure to and from adjacent buildings.

(c) **Provide for continuity of operations.** Protect the organization's ongoing mission, production, or operating capability. Minimize undue loss of operations and business-related revenue due to fire-related damage.

(d) **Limit the environmental impact of the fire.**

5.3   Allied Goals.

Securing life, property, operations, and the environment from damage or harm due to fire will often require the implementation of fire protection measures for the control or management of fire. The implementation of such fire protection measures can, however, impose a potential for nonfire damage, which might need to be considered. Allied goals of fire safety that might need to be developed and addressed in a performance-based design include, but are not limited to, the following:

(a) Provide for historic preservation by identifying a historic building's character-defining spaces, features, and finishes so that the implementation and the intended or unintended operation of fire protection measures do not result in damage or loss.

(b) Provide for protection of the environment by identifying the chemical or biological impact of the installation and the intended or unintended operation of fire protection measures on the natural environment.

5.4

The engineer must understand that there might be competing goals. Tenants and architects seek to maximize design flexibility, form, and function and still seek a fire-safe environment. Maintenance personnel desire fire protection systems that are easy to understand and maintain. A contractor's goals include ease of construction, and each stakeholder might have additional driving issues, such as time, money, or flexibility.

5.5

A goal is normally defined in broad terms by the stakeholders. There might also be supplemental goals or secondary goals. Table 5–1 provides examples of different goals, which the engineer should understand when conducting a performance-based design.

Table 5–1   Examples of Fire Safety Goals

Fundamental Goals

- Minimize fire-related injuries, and prevent loss of life

- Minimize fire-related damage to the building, its contents, and its historical features and attributes

- Minimize loss of operations and business-related revenue due to fire-related damage

- Limit the environmental impact of the fire and fire protection measures

Other Possible Goals

- Provide sufficient training and awareness to ensure the safety of the occupants from fire

- Reduce construction costs while maintaining adequate life safety measures

- Maximize design flexibility

- Minimize damage to historic building fabric

5.6   When undertaking a design based on a prescriptive-based code, the goal(s) might be embodied in an intent statement. When undertaking a design based on a performance-based code, the goal(s) might be embodied in an intent statement, embodied in an objective statement, or clearly stated.

5.7   Although the stakeholders might share the same global goals, the priority and relative weight might vary among stakeholders. Further differences might occur when defining objectives and performance criteria.

   5.7.1   Priorities should be based on the intended use of the building and its intended occupancy.

   5.7.2   Prioritization helps to clarify the intended use of potential fire protection measures, and it helps to identify those aspects of the fire protection analysis and design that require the most attention. (For example, if life safety is a high priority and property protection is a low priority, the fire protection analysis and design can focus on protecting the people until they reach a place of safety outside of the building, and it may not have to focus on protecting the building after the people have exited.)

   5.7.3   Although the stakeholders might identify only one or two goals as being important, all of the goals have to be addressed. For example, an industrial facility might be primarily concerned with process and property protection, yet it must also consider life safety and environmental issues. Here again, the process of prioritization makes the stakeholders consider a variety of potential concerns.

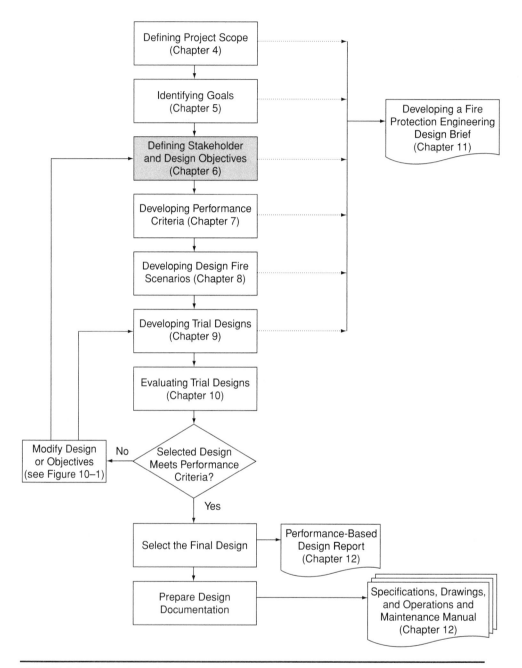

FIGURE 6–0 *Overview of the Performance-Based Design Process, Defining Stakeholder and Design Objectives*

6

Defining Stakeholder and Design Objectives

6.1 General

6.1.1 Once the fire protection goals have been established and agreed to, stakeholder objectives to meet the fire protection goals must be defined.

6.1.2 A stakeholder objective provides more detail than a fire protection goal, and it is often stated in terms of acceptable or sustainable loss or in terms of a desired (i.e., acceptable or tolerable) level of risk. Some stakeholders might, by virtue of experience or training, be able to state objectives in engineering terms that can serve as design objectives or performance criteria.

6.1.2.1 Stakeholder objectives might be stated broadly in terms of meeting one or more of the fundamental fire safety goals listed in 5.2. More specifically, objectives might be stated in terms of meeting the requirements of a specific code provision (prescriptive- or performance-based), of a specific insurance-related requirement, or in addition to a specific code or insurance provision or requirement.

6.1.2.2   Specific, project-related stakeholder objectives might reflect the maximum acceptable (tolerable) extent of injury to persons, damage to a building or its contents, damage to critical equipment or processes in the building, downtime or business interruption, risk ranking, or damage to the environment caused by fire or fire protection measures.

6.1.2.3   Stakeholder objectives might be expressed in different terms. For example, clients might express an acceptable loss as the maximum downtime or the amount of physical damage in dollars. Performance-based codes might state stakeholder objectives as an Objective, a Functional Statement, or a Performance Objective.

6.1.2.4   Regardless of the form in which they are stated, stakeholder objectives should be clear and agreed to by those involved because the engineer will later translate these objectives into numerical engineering values for design purposes.

6.1.2.5   Because most buildings contain ignition sources, fuels, and oxygen, there is always some likelihood that a fire could occur. Similarly, there is always some likelihood that a fire in an occupied building could result in injury or death, and a fire in any building will result in some property damage or business interruption. Therefore, the engineer must be clear, when communicating with other stakeholders, that it is not possible to create an entirely hazard- or risk-free environment.

6.2   Transforming Stakeholder Objectives into Design Objectives

6.2.1   To undertake an engineering analysis and design, stakeholder objectives must first be translated into values that can be quantified in fire protection engineering terms. These terms are the design objectives from which performance criteria can be developed. Quantification can be in deterministic or probabilistic terms. The stakeholder objective of confining flame damage to the compartment of origin could be translated into "not flashing over the room of fire origin" or "the probability of flashover is less than [a threshold value]."

6.2.2   The development of quantifiable design objectives should focus on the target(s), where the target is the building, compartment, process, or occupant that is being protected to meet a specific stakeholder objective. For example, targets might be the occupants of the facility (for a life safety fire protection objective), products or valuable equipment in a warehouse (for a property protection objective), a critical production or manufacturing process (for a continuity-of-operations objective), or water supply and wetlands (for an environmental objective).

6.2.3 Design objectives serve as the basis for the benchmark against which the predicted performance of a trial design will be evaluated using performance criteria expressed in engineering terms (Chapter 7). Appendix B shows examples of fire protection goals, stakeholder objectives, and design objectives.

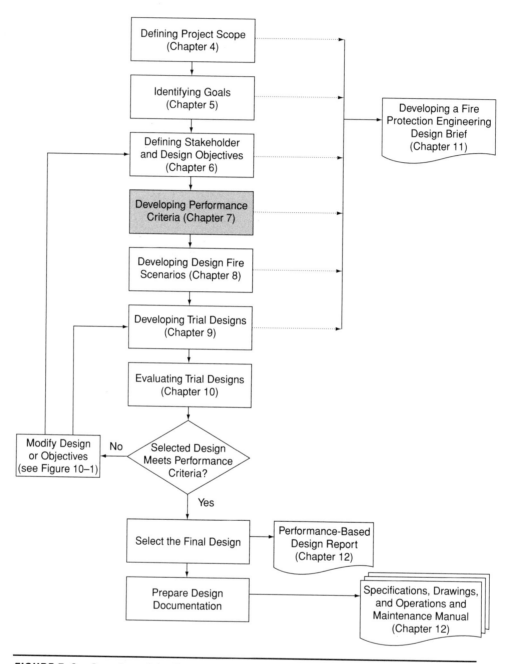

FIGURE 7–0 *Overview of the Performance-Based Design Process, Developing Performance Criteria*

7

Developing Performance Criteria

7.1 General

7.1.1 The next step in the performance-based design process requires the selection of performance criteria that will satisfy the design objectives and will be used to evaluate the trial designs. Performance criteria are threshold values, ranges of threshold values, or distributions that are used to develop and evaluate trial designs for a given design situation. Performance criteria might include temperatures of materials, gas temperatures, smoke concentration or obscuration levels, carboxyhemoglobin (COHb) levels, and radiant flux levels. Human response in terms of decision, reaction, and movement times varies over a range of values. To evaluate the adequacy of an egress system design with respect to human exposure criteria, it is necessary to select or assume values for calculation purposes. The rationale for assumptions regarding human behavior should be carefully documented. For example, performance criteria might include values for thermal radiation exposure (kW/m^2) or gas (air) temperature. Other types of performance criteria include concentration of toxic gases (ppm), distance of the smoke layer above the floor (m), visibility (m), or other measurable or calculable parameters. Examples of stakeholder objectives, design objectives, and performance criteria are provided in Appendix B.

7.1.2   More than one performance criterion might be required to adequately evaluate a design objective. In addition, more than one value might be required to adequately describe a performance criterion. For example, essential personnel might be required to delay their evacuation in order to secure an industrial process and, therefore, might experience conditions less tenable than nonessential personnel, who might evacuate sooner. These essential personnel might be provided with special equipment, training, or defend-in-place capability.

7.1.3   When defining performance criteria, note that it is impossible to achieve a completely hazard- or risk-free environment. Additionally, as the level of hazard or risk decreases, the costs associated with achieving those decreasing levels of risk typically increase.

7.1.4   In establishing performance criteria, the engineer should consider whether a method exists to evaluate a particular criterion. Evaluation methodologies are discussed in Chapters 9 and 10.

7.2   Setting Performance Criteria

Performance criteria are established from the design objectives developed by the engineer, which are based on the stakeholders' objectives. The design objectives are stated in engineering terms, but they lack the specificity required for comparison with the results of analytical hazard or risk assessment. The performance criteria must reflect the intent of the objectives and be quantitative measures of the consequences of fire that need to be avoided to fulfill stakeholder objectives. As such, performance criteria generally take the form of damage indicators. The means of preventing the damage need not be known at this stage, but a complete understanding of the acceptable limits of damage and injury must be well understood. Some performance criteria might also be set by applicable performance-based codes.

7.3   Establishing Specific Performance Criteria

7.3.1   Establishing specific numerical performance requirements for the universe of design situations is beyond the scope of this document.

However, the references in Appendix A might be useful for setting performance criteria. The list is not intended to contain all knowledge on a particular subject, only the fundamentals. Additional references should be used as appropriate. The following paragraphs identify areas in which performance criteria might be needed.

7.3.2 Life Safety Criteria Life safety criteria address survivability of persons exposed to fire and its products. The performance criteria might vary depending on the physical and mental conditions of the occupants and the length of time of expected exposure.

7.3.2.1 Thermal Effects

Analysis of thermal effects includes a threshold injury value and the exposure time needed to reach the threshold under the specific scenario being considered. Injury can result from exposure to thermal radiation[1,2,3] from either flames or heated gases. Radiation can also result in the ignition of clothing.

7.3.2.2 Toxicity

Toxic effects result from inhalation exposure to products of combustion.[3] The general effects on humans consist of reduced decision-making capacity and impaired motor activity, leading to incapacity or death. Even if victims escape, they might incur permanent damage. Analysis of these effects includes a threshold damage value and the exposure time needed to reach the threshold for the specific scenario being considered. Effects might vary depending on the age and health of those exposed. The increased temperature in a fire environment can result in rapid breathing and thus faster uptake of toxins.

7.3.2.3 Visibility

Visibility through smoke might affect the occupants' ability to safely exit a building.[3,4] The factors that affect visibility include the amount of particulate in the path of vision and the physiological effects on the eyes.[5] Low light levels might also affect occupants' ability to egress.

7.3.3 Non–Life Safety Criteria Non–life safety criteria address issues relating to damage thresholds for property. Damage thresholds may relate to thermal energy exposure, resulting in ignition or unacceptable damage. Thresholds might also consider exposure to smoke aerosols and particulate or corrosive combustion products. In some cases, unacceptable damage might result from small exposure levels.

7.3.3.1 Thermal Effects

Thermal effects might include melting, charring, deformation, or ignition. Considerations include the source of energy (e.g., convection, conduction, and radiation), the distance of the target from the source, the geometry of the source[1] and the target, the material characteristics of the target (e.g., conductivity, density, and heat capacity), and the ignition temperature of the target.[6,7,8] The surface area to mass ratio of the fuels involved is also a factor.

7.3.3.2   Fire Spread

The spread of fire by progressive ignition should be considered. Factors affecting fire spread include the geometry and orientation of the burning surfaces (horizontal versus vertical) as well as the surface area to mass ratio of the fuels involved.[9] Ventilation and airflow can increase or decrease fire spread.[8,9] Fire spread can also have an effect on life safety; rapid fire spread can impair occupant egress.

7.3.3.3   Smoke Damage

Smoke damage includes smoke aerosols and particulate or corrosive combustion products.[7] The damage threshold will depend on the sensitivity of the target to damage. Some works of art, such as paintings, have low thresholds, whereas others, such as statuary, might tolerate more smoke. Many targets, such as electronics, are sensitive to corrosive products at low levels.

7.3.3.4   Fire Barrier Damage and Structural Integrity

The loss of fire barriers can result in damage from the extension of heat and smoke. Opening protection operations and penetrations are factors. Minimal acceptable performance in terms of the amount of potential for extension will depend on the sensitivity of the target to heat and smoke. Structural collapse[10,11,12] is an issue in both life safety and property protection. The stability of a structure is important for occupants during the time necessary for egress and for emergency responders during rescue and suppression activities.

7.3.3.5   Damage to Exposed Properties

Performance criteria might need to be developed to prevent or limit damage or fire spread to exposed properties. The mechanism of damage can be heat or smoke. Separation distance, material flammability characteristics, and geometry are important considerations. Clearance of fire-prone materials and structures from wildlands might also be a criterion in some areas.

7.3.3.6   Damage to the Environment

Performance criteria, by limiting the effluent associated with fire suppression systems and fire-fighting operations or limiting the release of combustion products, might need to be developed to protect the environment.

References Cited

1. *Engineering Guide to Assessing Flame Radiation to External Targets from Liquid Pool Fires*, Society of Fire Protection Engineers, Bethesda, MD, 1999.

2. *Engineering Guide to Predicting 1st and 2nd Degree Skin Burns*, Society of Fire Protection Engineers, Bethesda, MD, 2000.

3. Purser, D. A. "Toxicity Assessment of Combustion Products," *The SFPE Handbook of Fire Protection Engineering*, National Fire Protection Association, Quincy, MA, 1995.

4. Bryan, J. L. "Behavioral Response to Fire and Smoke," *The SFPE Handbook of Fire Protection Engineering*, National Fire Protection Association, Quincy, MA, 1995.

5. Mulholland, G. W. "Smoke Production and Properties," *The SFPE Handbook of Fire Protection Engineering*, National Fire Protection Association, Quincy, MA, 1995.

6. Kanury, A. M. "Flaming Ignition of Solid Fuels," *The SFPE Handbook of Fire Protection Engineering*, National Fire Protection Association, Quincy, MA, 1995.

7. Tewarson, A. "Generation of Heat and Chemical Compounds in Fires," *The SFPE Handbook of Fire Protection Engineering*, National Fire Protection Association, Quincy, MA, 1995.

8. Drysdale, D. D. *An Introduction to Fire Dynamics*, 2nd Ed., John Wiley & Sons, Chichester, UK, 1999.

9. Quintiere, J. Q. "Surface Flame Spread," *The SFPE Handbook of Fire Protection Engineering*, National Fire Protection Association, Quincy, MA, 1995.

10. Milke, J. "Analytical Methods for Determining Fire Resistance of Steel Members," *The SFPE Handbook of Fire Protection Engineering*, National Fire Protection Association, Quincy, MA, 1995.

11. Fleishmann, C. "Analytical Methods for Determining Fire Resistance of Concrete Members," *The SFPE Handbook of Fire Protection Engineering*, National Fire Protection Association, Quincy, MA, 1995.

12. White, R. "Analytical Methods for Determining Fire Resistance of Timber Members," *The SFPE Handbook of Fire Protection Engineering*, National Fire Protection Association, Quincy, MA, 1995.

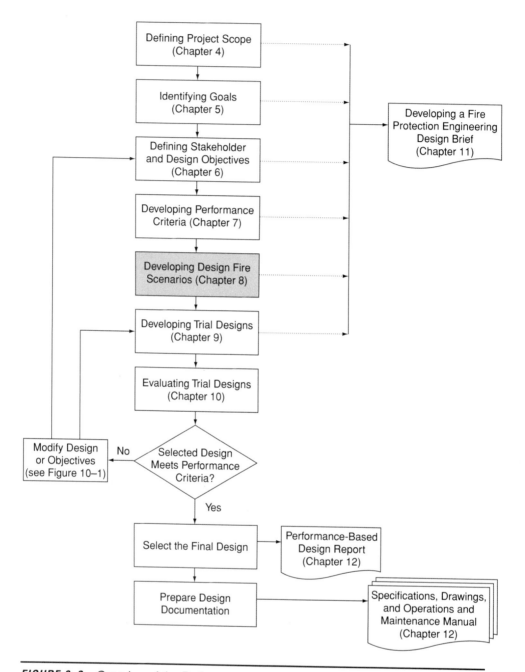

FIGURE 8–0 *Overview of the Performance-Based Design Process, Developing Design Fire Scenarios*

8

Developing Design Fire Scenarios

8.1 General

Once the performance criteria have been established, the engineer needs to focus on the development and analysis of design alternatives to meet these criteria. The first part of analyzing design alternatives is the consideration of possible fire scenarios, which are then filtered into selected design fire scenarios. Once design fire scenarios are established, then trial designs can be developed (see Chapter 9) and evaluated (see Chapter 10) to determine whether they meet the performance criteria for every design fire scenario. Figure 8–1 illustrates this process.

The process of identifying possible fire scenarios and developing them into design fire scenarios consists of the following steps:

- Consider possible fire scenarios (Section 8.2).
- Define the design fire scenarios, a subset of the possible fire scenarios (Section 8.4).
- Quantify the design fire scenarios.

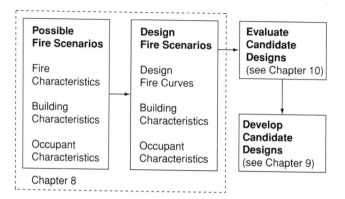

FIGURE 8–1   *Identifying, Developing, and Evaluating Fire Scenarios*

8.2   Identifying Possible Fire Scenarios

8.2.1   General

Fire scenarios describe factors critical to the outcome of fires, such as fire protection features, ignition sources, the nature and configuration of the fuel, ventilation, characteristics and locations of occupants, and conditions of the supporting structure and other equipment. A fire scenario might also include circumstances preceding ignition to the extent that they are necessary to help describe subsequent events.

A fire scenario represents one of a set of fire conditions that are thought to be threatening to a building, its occupants, and its contents. This description should therefore address the state of the building, its contents, and its occupants at the time of the fire.

Possible fire scenarios are the basis for design fire scenarios, which in turn are used to evaluate trial designs.

8.2.2   For a given possible fire scenario, many factors might affect fire development. These different factors might include the following:

- Form of ignition source
- Type of fuel first ignited
- Location of fire
- Effects of compartment geometry
- The initial status of doors and windows (whether open or closed) and the time in the possible fire scenario at which that status changed

- Ventilation, whether natural (e.g., doors and windows) or mechanical (e.g., HVAC)
- Type of construction and lining materials
- Form of intervention (e.g., occupants, fire suppression systems, and the fire department)

See Appendix D for an example of a possible fire scenario.

8.2.3 Possible Fire Scenario Characteristics

8.2.3.1   Prior to developing a possible fire scenario, the engineer should gather (or assume; see 8.2.3.2) various prefire characteristics of the specific building and the building's occupants. This information affects the chance of fire occurring, the ways it develops and spreads, and its potential to cause damage to the occupants, structure, and contents. This information will also be used as input variables in Chapter 9. Each scenario should define three components—fire characteristics, building characteristics, and occupant characteristics.

8.2.3.2   During design fire scenario development, some of the information might be unavailable. Therefore, the engineer might need to make explicit assumptions to address such points in the analysis. A sensitivity analysis (see 10.5.6.2) should be done to determine the assumptions that are likely to be influential.

The following provides an overview of the information that should be included when characterizing possible design fire scenarios. Additional information is provided in Section 8.5.

8.2.3.3   Building Characteristics

Building characteristics describe the physical features, contents, and ambient environment of the building. They can affect the evacuation of occupants, growth, and spread of a fire as well as the movement of combustion products. Therefore, building characteristics need to be detailed when developing fire scenarios.[1,2] Building characteristics include the following:

- Architectural features
- Structural components
- Fire protection systems
- Building services/processes
- Operational characteristics
- Fire department response characteristics
- Environmental factors

For further information on building characteristics, see Section 8.5.2.

8.2.3.4 Occupant Characteristics

Occupant characteristics need to be defined in order to determine occupants' abilities to respond and evacuate during an emergency. Thus, the following should be considered in the development of each possible fire scenario.[1,2]

- Number of occupants
- Distribution throughout the building
- Alertness (i.e., sleeping or awake)
- Commitment
- Focal point
- Physical and mental capabilities
- Role
- Familiarity
- Social affiliation
- Physical and physiological condition

See Section 8.5.3 for additional information.

The engineer should distinguish between factors that might be directly relevant to occupant behavior and escape (e.g., mobility) and factors that might be indirect indicators of direct factors (e.g., age and sex).

8.2.3.5 Fire Characteristics

Fire characteristics describe the history of a fire scenario, which includes the following:

- Ignition sources
- Growth
- Flashover
- Full development
- Extinction

The engineer should perform a thorough review of all potential and typical fuel packages and ignition sources for the building. It is often difficult to obtain specific information about building contents (e.g., furniture and stored materials) during the design stages of a project; however, every attempt should be made to understand what combustibles will be in the building. If assumptions are made regarding specific burning characteristics of materials, they should be documented and incorporated into final project specifications (such as architectural furniture specifications, if necessary). For additional information, see Section 8.5.4.

8.3 Tools Used to Identify Possible Fire Scenarios

A number of analysis techniques are available that can be used to identify possible fire scenarios; some of these techniques are described in the following sections.

8.3.1 Failure Modes and Effects Analysis (FMEA)[3]

Failure Modes and Effects Analysis is used to study systematically possible failure modes of individual components and the results of each failure, either on the system in general or on other components of the system.[4,5] FMEA originated in the aerospace industry and is used extensively in nuclear applications. However, the technique can also be generalized to assist in the development of possible fire scenarios in a variety of design applications.

A typical FMEA analysis first identifies the items to be studied. Next, the failure modes for each item are listed. A typical list of failure modes might include spark/arc, overheat, leak, rupture, loss of electrical power, or improper control signal input.

For each failure mode, a cause of the failure is then determined. For example, rupture might be caused by overpressurization, physical damage, or poor workmanship. The analysis then continues to identify the possible effects of each failure. The effects of a rupture obviously depend on what is released. If the material released is a combustible or flammable liquid, the effects might vary depending on a number of factors. For example, in the absence of an ignition source, the release might be only a pollution or cleanup problem; however, if a competent ignition source is present, a fire might result. If the rupture is caused by overpressurization during an existing fire, the effect could be a substantially increased growth rate and spread of the fire.

In the process of carrying out an FMEA, the engineer could assign some rank or criticality to the particular set of failure modes and effects being studied. The criticality value might consider not only the seriousness of the effect but also the likelihood that the particular failure mode might occur. Events that have a low frequency of occurrence but a serious effect might be more critical than events that occur frequently but are of little consequence (unless those events affect the system in such a way as to create another failure mode that leads to a greater effect).

8.3.2 Failure Analysis[3]

Failure analysis as used in engineering is the study of failures to determine the mechanisms by which the failures occurred and to develop prevention strategies. The American Society of Civil Engineers (ASCE)

Technical Council on Forensic Engineering defines failure as "an unacceptable difference between expected and observed performance."[6] Applying that concept to fire protection, a *fire failure* might be considered an event resulting in unacceptable deaths, injuries, or other losses.

Traditionally, failure analysis has been carried out after a failure occurred. Failure analysis can also be a part of the analysis and design process, and it can be used to anticipate failures that might compromise the expected performance of a fire protection design. Failure analysis applied during the analysis and design process is used to identify causes of potential failure, identify contributing factors, and evaluate expected system performance. Failures of fire protection systems might result from design, material, manufacturing or construction, and maintenance.

The analytical tools described elsewhere in this section are useful for identifying possible failure modes for fire safety design. See Appendix D for a sample application of failure analysis.

Failure analysis, then, can be an important tool for developing reliable fire protection designs. Failures are much easier to prevent before they occur than they are to control after they happen.

8.3.3   "What If?" Analysis[3]

This simplified technique involves asking what happens if a particular failure or event occurs. The answer will be an opinion based on the available knowledge of the stakeholders answering the question. The process can be enhanced by brainstorming among multiple stakeholders.

8.3.4   Historical Data, Manuals, and Checklists[3]

As a starting point for the analysis of an existing facility, historical fire incident data for that facility and associated processes or equipment can be reviewed for past scenarios. When a new project is involved, similar occupancy-, process-, or equipment-related fire statistics for the general population could be studied to identify common fire causes as well as other information. However, the scarcity of the data needed for assigning probabilities might limit this approach for conducting performance-based design for a specific project.

Operational manuals and checklists for processes or equipment could be studied to identify potential fire causes in possible fire scenarios. The engineer could look for warnings, cautions, and operational sequences that could lead to a fire if not followed. But, manuals and checklists might not identify all the possible fire causes or loss scenarios; they should be reviewed in order to gain a general understanding, which is used as a basis for asking questions of either a stakeholder or other resource.

8.3.5 Relevant Statistical Data[3]

When developing scenarios, it might be useful to review statistical data. Although it might not be appropriate to rely entirely on statistical information on fires in similar occupancies with similar fuel loads, past fire histories can provide some useful input, identifying potential failures of various items and frequencies and severities of fires. Reviewing statistical data, however, will not identify all foreseeable fire events.

Fire experience data can also help define high-challenge scenarios. High-challenge scenarios can be defined as those with high historical severity, such as high rates of death per fire. See Appendix C for additional information on the use of statistical data to choose possible fire scenarios.

8.3.6 Other Analysis Methods

Several other analysis tools exist that can be used, such as engineering checklists, hazard indices, hazards and operability studies (HAZOPS), preliminary hazards analysis (PHA), fault tree analysis (FTA), event tree analysis (ETA), cause consequence analysis, and reliability analysis.

8.3.7 Stakeholder input can be utilized to establish building characteristics based on plans. Occupant characteristics and fire characteristics can be based on the expected use of the structure. In some cases, the stakeholders can provide guidance in the development of fire scenarios for a building based on judgments of potential fuels or hazards or a survey of conditions.

8.4 Identifying Design Fire Scenarios

Given the large number of possible fire scenarios for a given performance-based design project, it is usually necessary to reduce the possible fire scenario population to a manageable number of design fire scenarios for evaluating trial designs. Generally, possible fire scenarios can be filtered into design fire scenarios using the engineer's judgment on what fires will bound the potential hazards. If calculations are necessary, two general approaches for accomplishing this are available—probabilistic and deterministic. Each design fire scenario (which is highly specific to support the hazard analysis calculation) is part of a scenario group (which is more general to support the frequency calculation) and is meant to be representative of that group. The scenario groups must collectively include all potential scenarios so that a valid risk measure can be calculated.

8.4.1 Probabilistic Approaches

A probabilistic approach typically deals with the statistical likelihood that a fire will occur and the outcome if a fire does occur. The decision whether to select a given possible fire scenario as a design fire scenario is based on grouping scenarios that are similar. Grouped sets of scenarios can potentially be eliminated from further evaluation if the stakeholders agree that the risk or outcome is acceptable. A probabilistic approach could use the following as sources of data:

8.4.1.1 Statistics and Historical Information

8.4.1.1.1 *Fire Statistics.* Fire statistics include statistics that identify the most likely areas of ignition, items first ignited, and the likelihood of spread beyond the room of fire origin. Fire statistics might be national or local, but they usually are applicable to a specific occupancy classification and type of building.[7] If a large database is available, assigning values for the frequency of occurrence of scenarios and establishing a relative ranking might also be possible. This process can result in a detailed risk analysis.

8.4.1.1.2 *Past History.* Past history includes historical data from fires in a particular existing building or group of buildings or in similar types of equipment, contents, and other items.

8.4.1.1.3 *Fire Frequencies.* The *fire frequency* is the number of times a fire occurs within a specified time interval. Fire frequency is usually measured in fires per year or events per year. Thus, if the fire frequency for a multiroom fire is stated to be 5×10^{-4} fires/year, the expectation is that a fire involving two or more rooms occurs every 2,000 years. In practice, the two-room fire might occur after this period, or it could occur next week.

8.4.1.1.4 *Fire Initiation Frequency.* The fire initiation frequency might be based on the building floor area[2,8] or use. See Appendix D for an example using building–floor–area fire initiation frequencies.

8.4.1.1.5 *Probabilities.* The likelihood that a specific outcome can occur can be assigned a numerical value between zero and one. Zero indicates that the event cannot occur; one indicates that the event is certain to occur. The probability of a specific outcome, *A*, can be estimated using the following formula:

$$P_A = \frac{n_A}{N}$$

where n_A is the number of ways that outcome *A* can occur, and *N* is the total possible outcomes.[9]

A *conditional probability* involves two defined events—events *A* and *X*. $P[A|X]$ represents the conditional probability of event *A* given event *X;* this is the probability of *A* occurring given that event *X* has already taken place.[10]

Typically, for fire protection analysis, the outcome is either pass or fail. The probability that a protective feature, given that a fire has occurred, would fail to control the fire could be P_{fail}. The probability of successful fire control given that a fire has occurred could be $P_{success}$. Because these probabilities require an event prior to their occurrence, they are referred to as *conditional probabilities*. References[10,11] are available for further information.

8.4.1.2   Hazard/Failure Analysis

This is a systematic analysis of the modes of failure and the fire hazards created for the specific occupancy. This analysis occurs more often for industrial premises where HAZOP, cause consequence, Failure Modes & Effects Analysis (FMEA), event trees, and fault trees can be used to develop scenarios.

8.4.1.3   System Availability and Reliability

8.4.1.3.1   *Fire Protection System Availability.* Fire protection systems might not always be operational. When establishing a fire frequency estimate, the concept of availability should be addressed. Availability is the availability of an entity (e.g., a fire protection system) to be in a state to perform a required function under given conditions at a given instant in time.[12] For fire protection purposes, the *given instant in time* is when a fire starts. If a fire detection system has been inadvertently left out of service after the completion of maintenance and a fire occurs, the detection system is, by definition, unavailable. Probabilities can be developed for the evaluation of system availability.

8.4.1.3.2   *Fire Protection System Reliability.* Fire protection systems do not always perform as designed or intended. When they fail, the results can be catastrophic. Thus, the reliability of fire protection systems should be understood and part of the design fire scenario selection. Reliability is a measure of the system's ability to perform as designed or intended.

Sometimes, the availability and reliability are reported or derived as a composite value. When this happens, the reference should be explicit in presenting this approach.

8.4.1.3.3   *System Availability and Reliability in Existing Building Analyses.* When developing design fire scenarios for the performance-based analysis of existing buildings, the availability and reliability of all fire protection systems, passive or active, in place at the time of the analysis should be considered. Such analysis is especially important when a design fire scenario relies on the successful operation of an installed system (see Section 8.5.2). For example, if an existing building has automatic fire detection and notification systems installed throughout, certain assumptions might be made during scenario development about the activation of the detection system and notification of building occupants. But, these assumptions should consider the probability that the system will be available (e.g., no detectors missing, no zones in trouble or not shut down) and that the system will operate as expected (e.g., respond as expected to the appropriate fire signature, provide alarm signaling that can be readily heard by building occupants and recognized as a fire alarm signal) (see also Section 8.5.3). Similar analysis might be appropriate for passive systems that might have been compromised at some point during occupancy of the building (e.g., holes in fire or smoke barriers, damaged or missing door-closing devices). Analysis of system availability and system reliability is not required when existing fire protection systems are not considered during the scenario development process. At other times, the system availability and system reliability will be addressed as part of trial design evaluations (see Chapter 10).

8.4.1.4   Risk

Risk is the distribution of scenario consequences and their frequencies of occurrence. Consequences might include occupant death, monetary loss, business interruption, or environmental damage; these are the consequences associated with an initiating event for a scenario of interest. The frequency of occurrence might be an estimate of how often the initiating event for a scenario might occur.

When considering risk as the basis for cost/benefit analysis, it might be possible to convert risk into equivalent dollars, which in turn allows different types of risks (e.g., death or injury, monetary loss, business interruption, or environmental damage) to be compared. This can be extremely difficult, however, due to difficulties in establishing a socially acceptable cost for a life lost. See Appendices D and E for sample applications of this concept.

8.4.2   Deterministic Approach

This approach relies on analysis or judgement based on physics and chemistry or correlations developed from testing to predict the outcome of a fire.

Fire scenarios can be evaluated for use as design fire scenarios by estimating whether a given scenario can produce effects that exceed the performance criteria. Those scenarios that do not exceed the performance criteria or are bounded by other design fire scenarios do not need to be used to evaluate trial designs.

In a deterministic analysis, one or more possible fire scenarios can be developed as design fire scenarios that are representative of potential worst credible fires in a particular building. They might include smoldering and rapidly developing flaming fires. The central challenge in scenario selection is finding a manageable number of fire scenarios that are sufficiently diverse and representative; if the design is safe for those scenarios, then it should be safe for all scenarios, except for those specifically excluded as too unrealistically severe or too unlikely to be fair tests of the design.

For deterministic design, the frequency of the possible fire scenario does not need to be evaluated. Each design fire scenario should be evaluated separately; multiple scenarios cannot be combined.

The analytical models and methods discussed in Appendix F can be used to determine if the scenarios would result in unacceptable outcomes.

Implied risk represents the unstated assumptions that the stakeholders have tacitly decided are an acceptable risk. Every trial design is intended to successfully prevent the fire from exceeding the performance criteria. Thus, all fires that might occur in a facility while the fire protection systems are functional and that do not exceed the severity of the design fire curve will be limited to less than the performance criteria. Those fires that develop more quickly than the design fire curve, have a higher heat release rate, or otherwise exceed the severity of the design fire might exceed the performance criteria. This potential illustrates the importance of selecting an appropriate design fire scenario.

8.5 Characterizing Design Fire Scenarios

8.5.1 General

This section provides information to assist in quantifying building characteristics, occupant characteristics, and design fire curves, which are used to characterize design fire scenarios, and it also provides information that might be necessary for developing possible fire scenarios.

8.5.1.1   The significant aspects of the design fire scenarios should be quantified. Often, neither the resources nor the data, such as ignition energy or heat release rate for each fuel package, are available to quantify every aspect of a design fire scenario. In these cases, the detailed analysis and quantification should be limited to the most significant aspects, which might include a range of different fire types (including smoldering fires), fire growth rates, or compartment ventilation rates.[2]

A design fire scenario that is highly improbable and too conservative can lead to an uneconomic building design, which might cause the building not to be built or be functional. On the other hand, a design fire scenario developed using a nonconservative approach (e.g., a long incipient phase or a slow rate of fire growth) could lead to a building design with an unacceptably high risk to occupants.

Design fire scenarios are not a description of how the majority of real fires in the building might behave. They are used to develop and test a trial design for robustness and, therefore, should present a conservative approach for analysis and determination of required fire safety measures.

8.5.1.2   Because design fire scenarios represent important input into any fire safety design, the stakeholders should agree to them as soon as possible in the design process.

The development of a design fire scenario might be a combination of hazard analysis and risk analysis. The hazard analysis identifies potential ignition sources, fuels, and fire development. The risk analysis might include the indicated hazard analysis while also noting the likelihood of the occurrence, either quantitatively or qualitatively, and the severity of the outcomes.

The design fire scenario should be described in sufficient detail to allow quantification of the scenario during scenario development and the evaluation of trial designs (see Chapter 10).

8.5.1.3   A portion of the characterization of the design fire scenario is establishing a design fire curve. The stages of fire growth and development of a design fire curve that should be reviewed are ignition, growth, flashover, decay, and burnout.[2]

It might not be necessary to quantify the aspects of each stage. For example, if in a performance-based design, the response of an alternative, automatic fire suppression system to standard sprinklers is being examined for equivalence, the design fire scenario might stop at the activation of the suppression system or at complete extinguishment.

For each design fire scenario, the agreement of all stakeholders on the design fire curve should be obtained.

8.5.2 Building Characteristics

8.5.2.1 Building characteristics describe the physical features, contents, and ambient internal and external environments of the building. They can affect the evacuation of occupants, the growth and spread of a fire, as well as the movement of combustion products.[1,2] Therefore, the items detailed below might need to be characterized when developing design fire scenarios.

8.5.2.2 The quantification of building characteristics might come from a prescriptive-based design option (if they are not part of the performance-based design) or other disciplines, such as mechanical engineering, electrical engineering, architecture, or interior design. In addition, they might be quantified as part of a trial design developed in Chapter 9. The extent to which building characteristics need to be quantified is a function of the level of analysis (see 10.2). For a subsystem-level design, only the factors pertinent to the design will need to be quantified. However, for a building performance-level design, all of the building characteristics will likely require quantification.

8.5.2.3 Architectural Features

The architectural features and construction of the compartments of interest as well as interconnecting compartments that might be affected may need to be determined. This information includes the following:

- The area and geometry of the compartment(s), ceiling height, and ceiling configuration (e.g., slope and beams)
- Interior finish flammability and thermodynamic properties (e.g., thermal conductivity, specific heat, and density)
- Construction materials and properties of walls, partitions, floors, and ceilings
- Position, size, and quantity of openings or areas of low fire resistance in the external envelope that could provide ventilation (e.g., windows and doors)
- Configuration and location of hidden voids (e.g., ceilings, floors, walls, suspended ceilings, and raised floors)
- Number of stories above and below grade
- Location of the building on the site in relation to property lines and other buildings or fire hazards
- Interconnections between compartments
- Relationship of hazards to vulnerable points

8.5.2.4   Structural Components

The structural aspects of the building might need to be determined. The structural aspects that should be considered are as follows:

- Location and size of load-bearing elements
- Construction material and properties of structural elements (e.g., strength, thermal conductivity, specific heat, reinforcement, and characteristics of connections)
- Protection material characteristics (e.g., thickness, thermal conductivity, and specific heat)
- Design structural loads

8.5.2.5   Fire Load

The fire load of the building should be estimated. Statistical data is available on the fuel loads for various occupancy types.[12] The extent to which the fire load is expected to vary over the life of the performance-based design should be considered. Where there is expected to be limited control over the fuel load, a conservative estimate should be considered.

8.5.2.6   Egress Components

Egress components, such as location, capacity of egress routes, and remoteness should be determined.

8.5.2.7   Fire Protection Systems

For existing buildings, there might already be fire protection systems that need to be considered. In addition, some fire protection systems might have already been predetermined, regardless of the analysis. This might include the following:

- The type of detection system (e.g., smoke detection, UV/IR, and heat detection), if provided, and the characteristics of the detectors (e.g., location of detectors, type of detector, and RTI) should be determined.
- The type of alarm notification, if provided, (e.g., voice, sounders, and strobes) and minimum sound pressure levels should be determined.
- The type and characteristics of any suppression systems should be determined (e.g., type of suppression, discharge density or concentration, location of discharge devices, activation characteristics such as type of activation or sprinkler activation temperature, and RTI).

8.5.2.8   Building Services/Processes

The location and capacity of building services, equipment, and processes should be determined based on the following elements:

- The location, capacity, and characteristics of ventilation equipment (e.g., mechanical versus natural, continually operating, and winter and summer differences)

- Effects on the ambient environment
- Location and capacity of electrical distribution equipment
- Potential ignition sources
- Process manufacturing equipment, which might need evaluation for property damage or business interruption aspects

8.5.2.9   Operational Characteristics

The operational characteristics of the building, including expected occupancy times (typically a function of time of day and day of the week), should be determined.

8.5.2.10   Fire Department Response Characteristics

Considerations with regard to the fire department (e.g., response time of fire fighters, accessibility of fire appliances, fire fighter access within the building, equipment, and services) might need to be determined. The location, capability, and response time of the fire department should also be investigated. See Section 9.5.4 for information on quantifying the effects of fire department operations.

8.5.2.11   Environmental Factors

The environmental factors both inside and outside the compartment of fire origin should be determined. These might include the following:

- Interior ambient temperature and humidity range
- Exterior ambient temperature and humidity range
- Ambient sound levels (as affecting alarm audibility)
- Expected wind conditions (as creating pressure differentials or stack effect)

8.5.3   Occupant Characteristics

The following occupant characteristics must be considered.

8.5.3.1   Human Behavior[13]

Human psychology plays an important role in life safety. The possible actions that might be taken following alarm notification need to be considered. These actions include searching for and rescuing family, friends, or pets; collecting valuables or records; and shutting down processes. Other factors include group behavior and fear reactions.

8.5.3.2   Response Characteristics

The occupants' response characteristics should be estimated. The response characteristics are a function of the following basic occupant characteristics: sensibility, reactivity, mobility, and susceptibility. They should reflect a population's expected distribution of characteristics as appropriate to the use of the building.

The four basic characteristics—sensibility, reactivity, mobility, and susceptibility—are occupant characteristics of people in buildings. The characteristics are briefly described as follows:

- *Sensibility to Physical Cues.* The ability to sense the sounding of an alarm or discern and discriminate visual and olfactory cues in addition to auditory emanations from the fire itself.
- *Reactivity.* The ability to interpret cues correctly and take appropriate action. This could be a function of cognitive capacity, speed of instinctive reaction, or group dynamics. The possible consideration of the reliability or the likelihood of a wrong decision, such as the influence of familiarity with the premises on wayfinding.
- *Mobility.* Speed of movement. A factor determined by individual capabilities as well as crowding phenomena, such as arching at doorways.
- *Susceptibility to Products of Combustion.* Metabolism, lung capacity, pulmonary disease, allergies, or other physical limitations that might affect survivability in a fire environment.

Response characteristics might address the following factors that are components of these basic occupant characteristics:

- *Alertness.* Awake/asleep, which might be a function of the time of day
- *Responsiveness.* The ability to sense cues and react
- *Commitment.* The degree to which an occupant is committed to an activity underway before the alarm
- *Focal Point.* The point at which an occupant's attention might be focused (e.g., the front of the classroom, the stage, or the server in a business environment)
- *Physical and Mental Capabilities.* The ability to sense, respond, and react to cues. Physical and mental capabilities might also be related to age or disability
- *Role.* A factor that might determine whether the occupant will lead or follow others
- *Familiarity with the Building and Evacuation Procedures.* A factor that can depend on time spent in a building or participation in emergency training
- *Social Affiliation.* A factor that can affect the extent to which an occupant will act/react as an individual or as a member of a group
- *Condition.* The physiological and psychological condition of the occupants, which might vary over the course of the fire depending on the exposure to fire and combustion products

8.5.3.3   Evacuation Times[14,15,16,17]

Prediction of occupants' movements during egress is an essential part of meeting life safety goals. The essential issue here is the estimation or prediction of the time needed for egress. A number of factors affect egress time including the number of occupants, distribution throughout the building, response to the notification (or fire alarm), age, health, mental capacity, motivation, and state of wakefulness of the occupants. The availability and level of training of the occupants or the staff in the case of institutional, assembly, educational, or health care occupancies are also factors.

The physical design of the egress system also affects egress time. The physical factors include the width and number of pathways as well as the width, tread depth, riser height, and number of stairways. Changes in width of the egress paths or the intersection of multiple egress paths result in congestion and slowing of movement.

8.5.3.4   Once occupants decide to evacuate, methods[15] and computer models are available to estimate the flow time through egress paths. The nature of estimating human behavior makes it difficult to accurately quantify the times of key events for the broad population. Therefore, careful attention should be paid to the treatment of uncertainty (see Section 10.5).

8.5.4   Quantifying Design Fire Scenarios and Design Fire Curves

8.5.4.1   Fire Development

As already stated, design fire curves are a vital part of the technical or engineering manifestation of design fire scenarios.

Design fire curves are time based and usually establish a relationship between heat release rate and time.

Design fire curves have a number of events and stages of development, which are shown in Figure 8–2.[2]

The design fire curve is usually described for a typical room or area of fire origin. Sometimes, only the growth phase up to flashover is considered, depending on the design objectives and trial designs being considered.

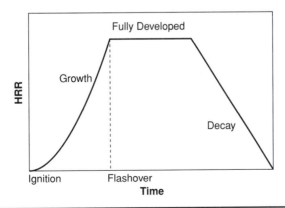

FIGURE 8–2   *Characteristics of a Design Fire*

Suppression might not form part of a performance-based design, and the fire may spread to adjoining rooms or spaces. In that case, design fire scenarios expressed in terms of heat release rate versus time are required for the subsequent rooms involved. This is particularly the case when fully developed fires (full room involvement) need to be examined for the effect on the performance of structural and separating (compartmentation) elements.

8.5.4.2   Design Fire Curve Stages

In any design fire curve that passes through a series of stages, it is often necessary to estimate the time of key stages. These stages might include ignition, preflashover, flashover, postflashover, and decay.

8.5.4.2.1   *Ignition Sources.*[3] Ignition requires the presence of three items—fuel, oxygen, and an ignition source. Ignition will not occur unless there is sufficient contact time with the source for the fuel to be raised to its ignition temperature given the energy of the ignition source. Because most buildings have fuel sources and oxygen, development of a design fire scenario might have to identify potential ignition sources. A deterministic approach might just assume ignition. In evaluating potential ignition sources, the temperature, energy, time, and area of contact with potential fuels should be considered.

The identification of ignition sources might be general. Ignition sources can be found throughout most parts of a building (e.g., electrical wiring and lighting) or might be located there at some point in the life cycle of the building (e.g., cutting and welding).

Having identified other aspects of a possible scenario, such as a particularly hazardous location, knowledge of ignition sources that might exist can help to establish the importance of the particular scenario.

An ignition with a low frequency of occurrence might result in the greatest loss, and it might need to be considered in the development of possible fire scenarios.

For design purposes, ignition is usually taken as the starting point or origin for the heat release rate ($\dot{Q}$) versus time (t) design fire curve.

Prefire situations are worth examining, at least in terms of fire prevention measures that could be taken in a building to minimize the risk of fires occurring.

However, for fire protection engineering purposes, ignition is typically the beginning of the design fire curve when self-sustaining combustion occurs.

Other initiators might include explosions, earthquakes, and earthquakes followed by a fire. In those buildings where an explosion potential exists due to the nature of the occupancy, or for those areas where earthquakes are a significant concern, appropriate consideration should be given to these initiator events.

8.5.4.2.2 *Growth (Preflashover) Stage.* In the growth stage, the worst credible fires might be rapidly developing flaming fires that have little or no incipient phase. However, designs should consider the possibility of scenarios and design fire scenarios with a relatively long smoldering phase that could cause fatalities or damage to critical targets even before established burning.

For smoldering fires, little data is available. The engineer should therefore choose carefully when developing any specific relationship for the design fire curve heat release rate.

For flaming fires, the growth rate is determined by characteristics of the initial and secondary fuels as well as the fire's potential to extend beyond the room of origin (i.e., extension potential).

8.5.4.2.3 *Initial Fuels.* Once the ignition sources are identified, then the characteristics of the initial fuels, or fuel packages, in the vicinity of these sources need to be evaluated.[18]

- *State.* A given fuel can come in various states (i.e., solid, liquid, or gas), and each state can have different combustion characteristics. A solid block of wood might be difficult to light with a match. Wood shavings might be easier to light, and if the wood were reduced to dust, it could be ignited by a spark and potentially create an explosion.

- *Type and Quantity of Fuel.* The development and duration of a fire depend on both the item that is burning and its state. Cellulosic-based materials and products burn differently than plastics or flammable liquids, producing different heat release rates, fire growth rates, and combustion products. The quantity of fuel and its form (surface area to mass ratio) will determine how long a fire could potentially burn.

- *Fuel Configuration.* The geometrical arrangement of the fuel will also influence the growth rate and heat release rate of a fire. A wood block will burn differently from a wood crib because of increased ventilation, surface area, and radiation feedback between the combustible materials.

- *Fuel Location.* The location of the fuel (e.g., against the wall, in the corner, in the open, and against the ceiling) will influence the development of the fire. Fires in the corner of a room or against a wall will typically grow faster than a fire located in the center of a room.[19]

- *Rate of Heat Release.* The amount of heat released per unit of time depends on the fuel's heat of combustion, the mass loss rate, the amount of incident heat flux, and the efficiency of combustion.[19] In addition to determining the time taken for the fire to have a significant effect on its surroundings, combustible items respond differently to various heating rates. The mass loss rate also directly relates to the production rate of smoke, toxic gases, and other combustion products.

- *Rate of Fire Growth.* The rate at which a fire grows is important because fire can be time dependent. Fires grow at various rates that are dependent on the type of fuel, its configuration, and the amount of ventilation. The faster a fire develops, the faster the temperature rises, and the faster the combustion products are produced.[19,20,21,22]

- *Production Rate of Combustion Products (Smoke, CO, CO_2).* Due to the different compositions of various fuels, as well as how they burn, the type and quantity of materials generated during combustion will be different. These products will include those that affect not only life safety, such as quantity of smoke, carbon monoxide, and carbon dioxide, but also those that affect property and business interruption, such as HCl, by damaging electronic equipment. Species production rates can be estimated using species yields, which are representative of the mass of species produced per mass of fuel loss.[20]

8.5.4.2.4   *Secondary Fuels.*[3] As a fire begins to grow in the initial fuel package, it can produce sufficient conductive, convective, and radiant energy to ignite adjacent fuel packages. Thus, in the development of a design fire scenario, the secondary fuel packages need to be identified and their characteristics defined in order to determine if they will become involved and allow the fire to grow, potentially to the point of full room involvement and even spread into adjacent compartments.

Ignition of secondary fuels can be by conduction, convection, radiation, or a combination of these.[2]

Conduction occurs when the heat in the fire plume is transported to the secondary fuel package. Important parameters regarding conduction include temperature of the reaction zone, conductivity, and the ignition temperature of the secondary fuel materials.[23,24]

Convection occurs when the heat in the fire plume transfers heat to the secondary fuel packages. Convection can also involve the carrying of embers that can cause piloted ignition. Important parameters regarding convection include the potential for flame impingement from the initial fuel source, temperature of the plume, heat transfer coefficient, and ignition characteristics of adjacent items.

Radiation is typically the main mode of heat transfer to adjacent fuel packages. Important parameters regarding radiation include the size of the flame, temperature, emissivity of the flame, absorptivity of the combustible surfaces, geometric viewing factor between the flames, the combustible surface of the adjacent fuel, and its ignition characteristics.

The factors that need to be identified with regards to these secondary fuel packages in addition to those outlined under the initial fuels section (8.5.4.2.3) include proximity to initial fuels, amount, and distribution.

8.5.4.2.5 *Extension Potential.*[3] The extension of a fire and its combustion products beyond the compartment or area of origin might also need to be considered. In addition to the fuel loading of the space, the construction features and layout of the compartment need to be analyzed in terms of how the fire and smoke can potentially extend beyond the original space.

A fire and its combustion products can extend to adjacent spaces in several ways—through openings via radiant energy or continuous fuel packages, through compartment walls via conduction, as well as through building services such as shafts and HVAC systems. The effects of building HVAC systems should also be considered as they can carry smoke and heat as well as toxic and corrosive products from one location to a location that might also be served by the same system.

Extension potential should be reviewed not only for fires extending within a building but also for fires extending over the external surfaces of the building or to adjacent buildings or structures.

8.5.4.2.6 *Target Locations.* When evaluating the expected development and spread of the fire, heat, and combustion products, the engineer should consider the location of the target items that correspond to stakeholder objectives.

8.5.4.3 For flaming fire scenarios, design fire curves might be developed based on the following items:

- Heat release data on specific, individual items likely to be in the building being designed, which can be aggregated to produce a heat release rate curve over time in a theoretical analysis

- Experimental heat release rate (HRR) data from items burned under a furniture calorimeter, in a corner burn test, or inferred from cone calorimeter data

- Full-scale test from mock-up sections of an actual building

- Generic curves for particular growth rates

- A fire growth model that can generate HRR data from fuel package data

For further information, see Section 8.5.4.4.

Based on an understanding of the concepts of preflashover fires identified above, the engineer should quantify the fire growth heat release rate as part of the design fire curve. This is usually tabulated or drawn as a curve, and it serves as input to analytical equations or fire models to determine key fire safety parameters and times to key stages.

8.5.4.3.1 *Flashover.* A key stage in fire safety is flashover (i.e., the point at which the heat release rate, temperature, smoke production, and smoke toxicity increase rapidly). The avoidance of flashover is often a critical fire protection engineering design objective for life safety and property protection.

The factors that affect whether flashover occurs in an enclosure include the following:[25]

- Surface area of the enclosure (A_t)
- The area of enclosure openings (A_v)
- The effective height of enclosure openings (h_v)
- Heat release rate
- Ventilation
- Thermal properties of compartment boundaries

Occupant evacuation of the room of fire origin before flashover is essential for life safety. Eliminating the chance of flashover analytically does not necessarily prevent fire extension or large-scale fire development. However, flashover does not always occur.

8.5.4.3.2 *Fully Developed (Steady or Postflashover) Fires.* In some design fire scenarios, the design fire might not reach flashover due to the fire occurring in a large space or limitations on available fuel or combustion air. The fire will reach a peak or steady heat release rate. This peak burning period will last for some period of time, which might be short or extended. The peak heat release rate is typically a factor of the fuel and the burning area.[26]

However, in some cases, fully developed fires will occur with total involvement of the fire enclosure, often leading to spread to other enclosures.

This fully developed part of the design fire curve will be used by the engineer in a design sense to determine the effect of the following items:

- Radiation through openings and potential ignition of adjoining buildings
- Failure of structural elements that might endanger life safety or cause unacceptable building or contents damage
- Fire spread to other enclosures through convected and radiated heat
- Failure of fire-resisting compartment elements designed to prevent fire spread to other parts of the building

The design fire scenario in the fully developed stage will be controlled by either the available ventilation or the available fuel. The engineer should calculate the heat release rate at ventilation control ($\dot{Q}_{vc}$) and fuel control ($\dot{Q}_{fc}$) and use the most appropriate of the two as the peak heat release rate for fully developed fires. This is illustrated in Figure 8–3.

The methods and data for establishing the fully developed heat release rate for this stage of the design fire curve might be found in various references.[25] Additional data sources are discussed in Section 8.5.4.4.

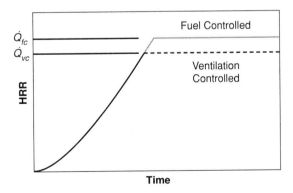

FIGURE 8–3   *Ventilation- and Fuel-Controlled Fires*

For designs involving compartmentation or structural fire resistance, the expected duration of the fire might need to be considered. To determine this value, it is necessary to determine or estimate a fuel load for design purposes. Fuel load is generally expressed in terms of mass of fuel per unit area.

8.5.4.3.3 *Decay and Extinction.* Fires will decay and eventually burn out after a period of time. Decay can be attributed to depletion of fuel load, lack of ventilation, or manual or automatic suppression systems extinguishing the fire. Correlations exist to predict the decay period for some scenarios.[27]

8.5.4.4 *Data Sources.* Several methods can be taken to define a design fire curve—using published data, using fire testing, or calculating a theoretical fire size. In a fire protection engineering design, it can be useful to consider more than one of these methods.

8.5.4.4.1 *Technical Literature.* When developing a design fire scenario and defining a design fire curve, the use of previously determined and reviewed data is an important tool. Various sources of heat release rates and associated data are available (see Appendix A).

8.5.4.4.2 *Fire Testing.* Fire testing can be used to collect heat release data that might be necessary to develop design fire curves. Many standard test methods are available for the collection of heat release data, ranging from small samples[28,29] to individual pieces of furniture[30,31] to complete room configurations.[32]

8.5.4.4.3 *Theoretical Methods.* Ideally, actual fire test data would be used when developing the design fire curves in a performance-based analysis; however, heat release and associated species generation data might not be available for the specific fuel packages and design fire scenarios under consideration, and resource limitations might preclude conducting fire testing. In these cases, the use of theoretical or nominal heat release rate and species generation curves might be necessary.

Empirical correlations for heat release rates have been developed from test data for a number of fuel packages, such as flammable liquid pool fires,[33] wood cribs,[34] upholstered furniture,[35] and electrical cable trays.[35] The application of these correlations requires knowledge of material properties, such as heat of combustion, size, and configuration of the commodity, and in some cases, information on ventilation rates.

The heat release rate during the growth phase of a fire can, in many cases, be generically represented by a time-dependent exponential function. One such approach, commonly referred to as a t^2 *fire,* is to represent the heat release rate as increasing proportionately to the square of the time since ignition.

In addition, several fire-modeling software packages, such as FPETool[36] and HAZARD I,[37] contain routines for estimating fire growth curves to represent the cumulative heat release rate due to the burning of multiple fuel packages. These routines require the input of fire growth curves for each fuel package considered and information regarding the spatial separation and physical arrangement of the fuels. One or more fuel packages are designated as being initially ignited, and the routine outputs the resulting fire growth curve from the subsequent ignition of the other fuel packages.

Caution must be exercised when utilizing a theoretical fire growth curve in a performance-based analysis because the resulting fire growth curves are approximations of the anticipated fire conditions. Appropriate conservatism and safety factors, such as selecting a larger than expected heat release rate, might need to be included in the analysis in order to compensate for the uncertainty associated with the theoretical computational methods for developing the fire growth curves (see Section 10.5).

Appendix F provides guidance on the selection of analytical methods.

8.6   The design fire scenarios will form the basis for evaluation (see Chapter 10) of the trial designs (see Chapter 9).

References Cited

1. "Fire Safety Engineering in Buildings," DD 240, British Standards Institute, London, 1997.

2. *Fire Engineering Guidelines,* Fire Code Reform Centre Limited, Sydney, NSW, Australia, 1996.

3. Custer, R. L. P., & Meacham, B. J. *Introduction to Performance-Based Fire Safety*, National Fire Protection Association, Quincy, MA, 1997.

4. Hensley, E., & Kumamoto, H. *Reliability Engineering and Risk Assessment*, Prentice-Hall, Englewood Cliffs, NJ, 1981.

5. Hammer, W. *Handbook of System and Product Safety*, Prentice-Hall, Englewood Cliffs, NJ, 1972.

6. Leonards, G. "Investigation of Structural Failures," *Journal of the Geotechnical Engineering Division*, American Society of Civil Engineers, 108, # GT2, February 1988.

7. Hall, J. "Use of Fire Incident Data and Statistics," *Fire Protection Handbook*, National Fire Protection Association, Quincy, MA, 1996.

8. Coutts, D. A. "Fire Risk Assessment Methodology Generic Event Tree Description (U)," Westinghouse Savannah River Company, Aiken, SC, 1994.

9. Wadsworth, H. M., Ed. "Summarization and Interpolation of Data," *Handbook of Statistical Methods for Engineers and Scientists*, McGraw-Hill Publishing, New York, NY, 1990.

10. Villemeur, A. *Reliability, Availability, Maintainability and Safety Assessment*, John Wiley & Sons, Chichester, UK, 1992.

11. Hall, J. "Probability Concepts," *The SFPE Handbook of Fire Protection Engineering*, 2nd Ed. National Fire Protection Association, Quincy, MA, 1995.

12. CIB, *A Conceptual Approach Towards a Probability Based Design Guide on Structural Fire Safety*, CIB W14 Workshop Report, Rotterdam, Netherlands, 1983.

13. Bryan, J. L. "Behavioral Response to Fire and Smoke," *The SFPE Handbook of Fire Protection Engineering*, National Fire Protection Association, Quincy, MA, 1995.

14. Pauls, J. "Movement of People," *The SFPE Handbook of Fire Protection Engineering*, National Fire Protection Association, Quincy, MA, 1995.

15. Nelson, H. E., & MacLennan, H. A. "Emergency Movement," *"The SFPE Handbook of Fire Protection Engineering*, National Fire Protection Association, Quincy, MA, 1995.

16. Shields, J., Ed. *Proceedings of the First International Symposium on Human Behavior in Fire*, University of Ulster, 1998.

17. Section 8, "Evacuation of Occupants," *Fire Protection Handbook*, National Fire Protection Association, Quincy, MA, 1996.

18. Bukowski, R. "Fire Hazard Analysis," *Fire Protection Handbook*, National Fire Protection Association, Quincy, MA, 1996.

19. Zukoski, E., Kubota, T., & Cetegen, B. "Entrainment in Fire Plumes," *Fire Safety Journal*, vol. 3, pp. 107–121, 1980.

20. Tewarson, A. "Generation of Heat and Chemical Compounds in Fires," *The SFPE Handbook of Fire Protection Engineering*, National Fire Protection Association, Quincy, MA, 1995.

21. McCaffrey, B. "Purely Buoyant Diffusion Flames: Some Experimental Results,"National Institute of Standards and Technology, NBSIR 79–1910, Gaithersburg, MD, 1979.

22. Thomas, P., Hinkley, P., Theobald, C., & Simms, D. "Investigations into the Flow of Hot Gasses in Roof Venting," Fire Research Technical Paper No. 7, HMSO, London, 1963.

23. Drysdale, D. D. *An Introduction to Fire Dynamics,* 2nd Ed., John Wiley & Sons, Chichester, UK, 1999.

24. NFPA 72, *National Fire Alarm Code,* National Fire Protection Association, Quincy, MA, 1999.

25. Walton, W. D., & Thomas, P. "Estimating Temperatures in Compartment Fires," *The SFPE Handbook of Fire Protection Engineering,* 2nd Ed., National Fire Protection Association, Quincy, MA, 1995.

26. Lie, T. T. "Fire Temperature-Time Relations," *The SFPE Handbook of Fire Protection Engineering,* 2nd Ed., National Fire Protection Association, Quincy, MA, 1995.

27. Evans, D. "Sprinkler Fire Suppression Algorithm for HAZARD," NISTIR 5254, National Institute of Standards and Technology, Gaithersburg, MD, 1993.

28. "Standard Test Method for Determining Material Ignition and Flame Spread," LIFT E1321, American Society for Testing and Materials, West Conshohocken, PA, 1997.

29. "Standard Test Method for Heat and Visible Smoke Release Rates for Materials and Products Using an Oxygen Consumption Calorimeter," E1354, American Society for Testing and Materials, West Conshohocken, PA, 1999.

30. "Standard Test Method for Fire Testing of Upholstered Seating Furniture," E1537, American Society for Testing and Materials, West Conshohocken, PA, 1999.

31. "Standard Test Method for Fire Testing of Mattresses," E1590, American Society for Testing and Materials, West Conshohocken, PA, 1999.

32. "Standard Guide for Room Fire Experiments," E603, American Society for Testing and Materials, West Conshohocken, PA, 1998.

33. Mudan, K. S., & Croce, P. A. "Fire Hazard Calculations for Large Open Hydrocarbon Fires," *The SFPE Handbook of Fire Protection Engineering,* 2nd Ed., National Fire Protection Association, Quincy, MA, 1995.

34. Babrauskas, V. "Burning Rates," *The SFPE Handbook of Fire Protection Engineering,* 2nd Ed., National Fire Protection Association, Quincy, MA, 1995.

35. Babrauskas, V., & Grayson, S. J., Eds., *Heat Release in Fires,* Elsevier Applied Science, New York, NY, 1992.

36. Deal, S. "Technical Reference Guide for FPETool Version 3.2," NISTIR 5486–1, Building and Fire Research Laboratory, National Institute of Standards and Technology, Gaithersburg, MD, 1995.

37. Peacock, R., Jones, W., Bukowski, R., & Forney, C. *Technical Reference Guide for the HAZARD I Fire Hazard Assessment Model, Version 1.1,* National Institute of Standards and Technology, Gaithersburg, MD, 1996.

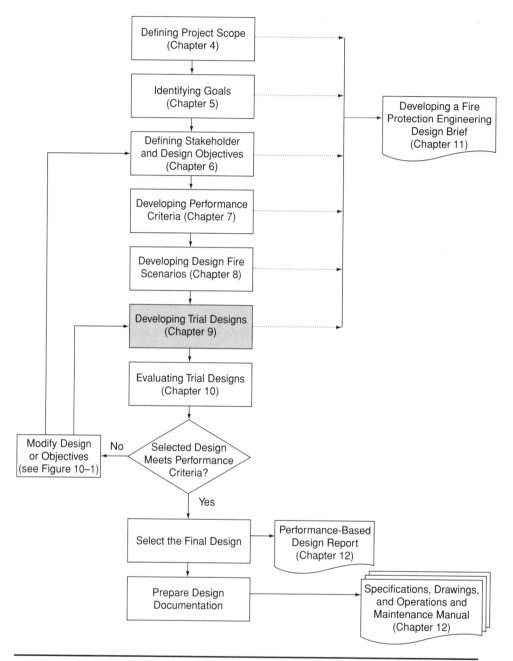

FIGURE 9–0 *Overview of the Performance-Based Design Process, Developing Trial Designs*

9

Developing Trial Designs

9.1 General

9.1.1 After the performance criteria have been established and the design fire scenarios have been determined, the next step is the development of trial designs. Design fire scenarios developed in Chapter 8 are used to test these trial designs, and the results of the analysis are evaluated (Chapter 10) using the performance criteria determined in Chapter 7.

9.1.2 When developing trial designs, the context in which the design will be evaluated should be considered. Trial designs can be evaluated at a subsystem level, which might involve a comparison with the provisions of a prescriptive-based design option, or on a system-performance or building-performance basis, which relies on an evaluation relative to established performance criteria.

9.1.3 Trial designs developed in the context of comparison evaluations might require a comparison of the performance of the design features of a prescriptive-based design option with the performance resulting from the trial design. Developing trial designs in this context might simply require selecting features similar to that of the prescriptive-based design option, but with enhanced capabilities or features. On the other hand, features might be selected that provide appropriate safeguards in a different manner. For example, in lieu of extinguishing a fire to mitigate a smoke hazard, venting could be used to purge the smoke hazard. Using prescribed features as a baseline for comparison, the evaluation could then demonstrate whether a trial design offers the same level of performance.

9.1.4   Trial designs evaluated using performance criteria require the selection and development of design features that should fulfill the performance criteria for the design fire scenarios under consideration. The design features being developed should consider the capabilities, reliability, costs, and maintenance requirements of the trial design.

9.1.5   The Fire Protection System

9.1.5.1   The level of fire safety in a building is a function of the interaction of each of the components, or subsystems, of the fire protection system.

9.1.5.2   The subsystems that are discussed in this guide are as follows:
1. Fire initiation and development
2. Spread, control, and management of smoke
3. Fire detection
4. Fire suppression
5. Occupant behavior and egress
6. Passive fire protection

9.1.5.3   Although each subsystem may be evaluated separately, the interactions between subsystems must be determined. For system-performance or building-performance evaluations, the interactions between the subsystems will need to be considered.

9.1.5.4   For any given design, not all of the subsystems described in this guide have to be used. The combination of subsystems used for any given design is a function of the performance criteria the design is intended to achieve.

9.1.6   Setting Performance Criteria

Various methods of achieving a design objective must be considered when setting performance criteria. These methods will use fire protection approaches such as suppression, detection, compartmentation, and material flammability control.

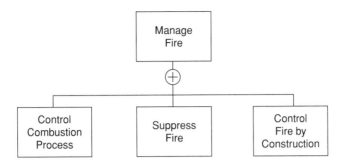

FIGURE 9–1   *The Fire Safety Concepts Tree*

9.1.6.1   NFPA 550,[1] the *Fire Safety Concepts Tree,* can assist with the identification of general approaches and methods for achieving a given design objective. NFPA 550 shows three methods of managing a fire—*control combustion process, suppress fire,* and *control fire by construction* (see Figure 9–1). The *or* gate is represented by a +, while *and* gates are represented by •. Because these options are separated by an *or* gate, any one of them may be used. Each of these methods is further subdivided until the lowest level of the tree is reached. The concepts contained on this lowest level are areas that should be considered when developing performance criteria. For concepts separated by an *and* gate, performance criteria must be developed and met for each concept. However, for concepts separated by *or* gates, only one of the performance criteria associated with the concepts needs to be met.

Some concepts might need to be broken down to lower levels than contained in NFPA 550. For example, *apply sufficient suppressant* might need to be further subdivided depending on what type of suppressant is used. Also, the time associated with agent delivery might need to be considered because the quantity of agent required could be a function of fire size, which is typically a function of time.

For example, consider a design objective of *limit the spread of flame to the compartment of origin.* In NFPA 550, this objective corresponds to the *Manage Fire* branch of the tree (see Figure 9–2). The *Manage Fire* branch may be met by either controlling the combustion process, fire suppression, or controlling the fire by construction. Following the tree to the lower levels will yield possible strategies, which can be developed into performance criteria associated with managing the fire.

9.1.6.2   In addition to using NFPA 550 to determine suitable performance criteria, the methods identified in Chapter 8 (Failure Modes

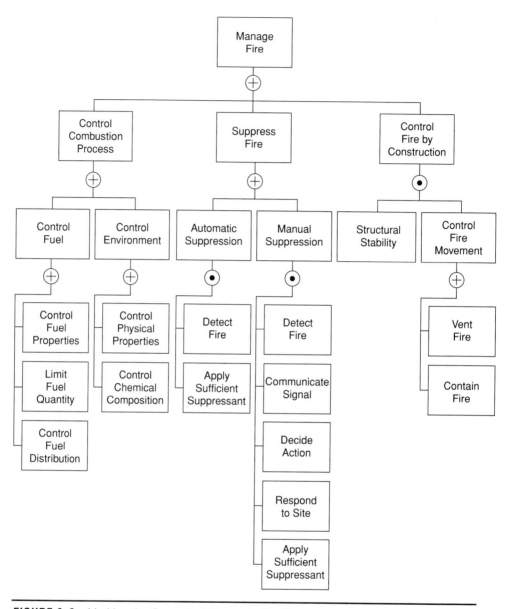

FIGURE 9–2   *Limiting the Spread of Flames to the Compartment of Origin*

and Effects Analysis, failure analysis, "What if?" analysis, event tree analysis) can be used to determine methods of achieving design objectives.

9.1.6.3   When a performance-based design needs to be *equivalent* to a mandated requirement, such as in a law, regulation, or code, determining what performance is intended by the mandated or prescriptive requirement will be necessary. For example, if a stakeholder wants automatic sprinklers or *equivalent,* the fire size at the time of sprinkler operation, the level, and the distribution of smoke and other combustion products must be quantified for the specific building and fire scenario in question. This establishes the baseline performance equivalency for alternative designs.

9.1.7 Options for Trial Designs

9.1.7.1   Several options from which to choose are available for any given fire protection system. The choice of which subsystems to include in trial designs is dependent on the performance criteria developed in Chapter 7.

9.1.7.2   Different trial designs should be developed using different combinations of these components. The design team should be involved in the development of the different combinations of components.

9.1.7.3   The options for trial designs are discussed in 9.2–9.7.

9.2 Fire Initiation and Development

9.2.1 Prevention of Ignition

9.2.1.1   *Objective.* To reduce the probability (or likelihood) that ignition will occur

9.2.1.2   Concepts
1. *Control of Sources.* Ignition sources are controlled in three ways.
   - *Removal.* Remove ignition sources from the area.
   - *Isolation.* Separate ignition sources from combustibles by physical space.
   - *Management.* Select heat-producing equipment that uses intrinsically safe designs.
2. *Control of Materials.* Selecting materials that are inherently resistant to ignition can control the propensity for ignition. Materials that have a high thermal inertia will be inherently less prone to ignition than materials with a low thermal inertia. For additional information on thermal inertia, see *The SFPE Handbook of Fire Protection Engineering.*[2] Also, materials that use fire retardant might be considered; however, the effect of time and wear on the fire retardant should be considered.
3. *Fire Safety Management.* In addition, good housekeeping practices might be used as a method of controlling ignition as well as the amount and type of combustibles that are present.

9.2.2   Control of Fire Development

9.2.2.1   *Objective.* To reduce the development rate of a fire and the related smoke and heat production

9.2.2.2   Concepts

1. *Contents Selection.* The rate of fire growth might be controlled by the selection and arrangement of contents used in the building. Some contents have a high rate of heat release, a factor that might be dependent on content configuration[3] (e.g., saw dust versus a block of wood). Materials with a smaller propensity for ignition can control fire growth and fire spread from object to object.

2. *Placement of Contents.* The separation distance between objects is related to the ease with which fire spreads from object to object.

3. *Interior Finish Selection.* Similar to contents, combustible interior finishes might cause the fire to spread faster, particularly when placed on vertical surfaces or on the ceiling.

4. *Compartment Geometry.* Large compartments with high ceilings tend to reduce the amount of radiant feedback to the fire and, therefore, decrease the rate of fire growth.

5. *Ventilation Control.* Fire development can be managed by controlling ventilation to the fire. However, when using ventilation control as a means of limiting fire growth, the effects on combustion gas temperatures, ignition of incomplete combustion gases remote from the fire, and carbon monoxide concentration should be considered.

6. *Suppression Systems.* Suppression systems can be provided to limit the amount of heat being released by the fire and the size to which it grows.

7. *Construction.* The rate of fire growth might be affected by the construction features and design.

9.3   Spread, Control, and Management of Smoke

9.3.1   *Objective.* To reduce the hazards resulting from smoke by limiting its production, controlling its movement, or reducing the amount of it

9.3.2   Concepts

1. *Material Control.* Materials in the building and its construction can be controlled to exclude those materials that produce large quantities of toxic smoke (e.g., plastics).

2. *Containment.* Smoke can be contained to its area or compartment of origin by using smoke control doors, smoke dampers, lobbies, or other items.

3. *Extraction.* Smoke can be removed from a compartment by natural or mechanical means. Large volume spaces, such as atria, might be protected with smoke extraction systems that maintain a smoke-free layer in which occupants can egress and fire fighters can mount fire attack efforts.

4. *Pressurization.* Pressure differentials can be created to direct the movement of smoke from high-pressure areas to low-pressure areas. This technique is commonly used to prevent smoke from getting into lobbies and stairways and is sometimes used to pressurize the floors above and below the fire floor and exhaust the fire floor, or to keep smoke on the level of origin. By dividing a building into smoke zones and creating a pressure differential across those zones, smoke movement through cracks in floors and through shafts can be minimized. Zoned smoke control systems typically require a supplemental detection system that can control in which zones positive and negative pressures are created.

5. *Suppression.* Application of a suppression system can reduce smoke development and limit smoke movement by limiting fire growth.

In this guide, smoke includes the fire products of combustion and the air that is entrained into the fire plume. The primary constituent of smoke is the air that is entrained. When considering mass or volumetric smoke production rates, the mass flow of combustion products (CO_2, H_2O, CO) are typically ignored because they constitute a small portion of the smoke.[4] Their production rates will be considered in Section 9.3.3.1, Species Concentration.

Several correlations are available[5,6,7] for flaming fires that yield smoke mass flow rate as a function of the height above the base of the fire. The design fire scenario forms the basis for the inputs for these correlations. Two of these references[4,6] might also be used to estimate the plume temperature as a function of height.

Detailed information on the design and evaluation of smoke management systems might be found in *Design of Smoke Management Systems*,[8] NFPA 92A[9] and NFPA 92B.[10]

Activation of a smoke management system typically occurs by either automatic detectors or manual means. The activation time can be determined in accordance with the detection and notification subsystem.

9.3.3 Other Considerations

9.3.3.1 Species Concentration

Estimation of species concentration in smoke requires information, such as the ratio of air to fuel in the combustion zone. Tewarson provides a discussion of species generation in fires in *The SFPE Handbook of Fire Protection Engineering*.[11] In the absence of data on species production rates, a conservative approach might be to attempt to avoid exposure of occupants to smoke.

9.3.3.2 Smoke Optical Density

The SFPE Handbook of Fire Protection Engineering provides a table of maximum specific optical densities for a variety of commonly used materials.[12] This data should be used with caution because it was developed based on small-scale tests and might break down for complex fires. Again, in the absence of data, a conservative approach might be to attempt to avoid exposure of occupants to smoke.

9.3.3.3 Upper-Layer Temperature and Depth

Calculation of the upper-layer temperature and depth is accomplished with calculations based on empirical or theoretical equations. These calculations might utilize a computer fire model or hand calculations.

9.3.3.4 Smoke Flow from Enclosures

The calculation of smoke flow from enclosures through openings, such as doors, windows, and ventilation ducts, might be accomplished with certain fire models. Several calculation techniques are contained in *The SFPE Handbook of Fire Protection Engineering*.[13]

9.4 Fire Detection and Notification

9.4.1 Fire Detection

9.4.1.1 *Objective.* To provide warning of a fire in order to notify occupants or emergency personnel, or to activate an active fire protection system, such as a smoke management or suppression system

9.4.1.2 Concepts

1. *Human Detection.* Humans can detect fires. However, they must be in the vicinity, be alert, and be capable of responding.

2. *Automatic Detection.* Electro-mechanical detectors can be provided that monitor a space for various types of fire signatures

(e.g., heat, smoke, and radiant energy). Automatic detectors might be connected to an alarm system to alert building occupants as well as the fire department.

9.4.1.3 Delays

Most detection systems have inherent delays, which could be either variable or fixed.[14] Variable delays represent either transport lags associated with combustion products reaching the detector or detector-response delays. The delay associated with the transport of combustion products is a function of the heat release rate of a fire, the ceiling height, and the radial distance of the detector from the fire. The delay associated with the detector response can be quantified if the detector response time index (RTI) is known. For example, the computer model DETACT accounts for detector response delays by requiring RTI as input data.[15]

In many cases, a quasisteady state is assumed, and the transport delays are ignored. However, the quasisteady assumption might not be appropriate when long transport delays are expected (e.g., in high ceiling spaces or when very rapid detection is desired).

Fixed delays are associated with system characteristics, such as polling system delays or alarm verification time. System delays can be ascertained from the manufacturer's data.

9.4.1.4 Estimation of Detector Operation Time

Computer modeling is typically used to estimate thermal detector operation time. These models typically require the detector response time index, temperature rating, and enclosure geometry, such as the ceiling height above the fire and the detector radial distance from the fire centerline, as input. *The SFPE Handbook of Fire Protection Engineering* provides information on estimating smoke detector response time.[16]

9.4.2 Notification

9.4.2.1 *Objective.* To notify the occupants of a building or the fire department that a fire has started and possibly to provide the location of the fire

9.4.2.2 Concepts

1. Notification systems might be either human or electro-mechanical (i.e., automatic). If the notification system relies on humans, then the reliability of the system should be considered as part of the design.

2. Audible alarms or visual cues might be initiated by a detection system to notify occupants in the area of fire origin that a fire has started.

3. Automatic detection system output can also be used to initiate notification of the fire department that a fire has started. If the detection system is intended to notify the fire department, an important consideration is whether the detection system output is sent to a central station or directly to the fire department. Criticality or importance of the protected facility might dictate where the detection system output is sent.

4. The process the fire department uses to receive and process a notification signal (i.e., alarm handling) might have an impact on its response to the fire.

5. Notification might also include information as to where the fire is (e.g., third-floor electric closet), once the fire department is on site. Notification of the fire's location can be performed by a detection system connected to an annunciator panel.

9.5   Fire Suppression

9.5.1   *Objective.* To extinguish or control a fire

9.5.2   **Concepts**

1. *Automatic Suppression Systems.* Automatic suppression systems require no human interaction. These are either stand-alone systems or ones that require a detection system to activate them. Suppression systems include sprinkler systems, gaseous suppression systems, and foam systems.

2. *Manual Suppression Systems.* Manual fire fighting includes fire-fighting actions taken by occupants, fire brigades, or fire departments. First-aid fire-fighting appliances (i.e., those intended for use by building occupants), such as fire extinguishers, hose reels, and sand buckets, can be used to begin fire suppression activities. However, unless the fire is small, these efforts are not necessarily expected to extinguish the fire. Therefore, additional suppression might be needed. This additional suppression might be provided by automatic suppression systems or by organized manual suppression systems, such as fire departments and fire brigades.

Suppression systems can be used to meet life safety and property protection objectives, and they can be used as a means of fire detection (e.g., a water flow switch on a sprinkler system).

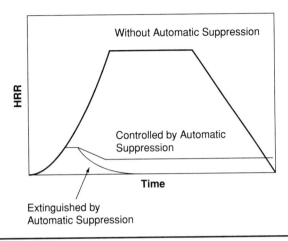

FIGURE 9–3   *Impact of Automatic Fire Suppression Systems*

9.5.3   Automatic Fire Suppression Systems

9.5.3.1   Impact of Automatic Fire Suppression

- When automatic fire suppression systems are installed, assuming they are activated when they are designed to do so, the design fire curve might be modified (see Figure 9–3).

The time at which the suppression system activates is dependent on the system design and particularly the mechanism used to activate it. Section 10.4.2.2 identifies a procedure for estimating the activation time of suppression systems.

An important concept is that of control versus extinguishment (complete suppression). Traditional sprinkler systems might only control fires (i.e., to prevent the heat release rate from increasing further). On the other hand, gaseous systems and some forms of recent sprinkler Early Suppression, Fast Response (ESFR) systems might fully extinguish the fire. This is illustrated in Figure 9–3.

9.5.3.1.1   *Delays.* As with detection systems, suppression systems might have delays associated with system operation. In addition to the delays described in Section 9.4.1.3, these suppression system delays will include delays associated with agent discharge, delays reaching an extinguishing concentration, and predischarge warning delays.

For wet pipe sprinkler systems, the discharge delay can typically be ignored. However, for dry pipe or deluge systems, the discharge delay needs to be accounted for. *The SFPE Handbook of Fire Protection*

Engineering provides a formula for estimating the operation time for a dry pipe valve.[17] However, as there is no established method to estimate the actual time required for water to travel from the valve to the open sprinklers,[18] a suitably conservative time should be assumed.

Software is commercially available[19] to estimate the discharge time for gaseous suppression systems.

9.5.4 Manual Fire-Fighting Operations

9.5.4.1 Concepts

1. *Notification.* The emergency response forces will require notification of a fire.

2. *Access.* The emergency response forces should have easy access for their equipment not only on site but also into the building and to the area of fire origin. The existence of gates and fences as well as secured buildings will delay the fire department's response to the area of origin. Access might also include protected stairways and corridors so that fire department personnel could get as close to the origin as possible.

3. *Water Supply.* The fire department should be provided with an adequate water supply on site (e.g., fire hydrants and cisterns) as well as within the building (e.g., standpipes and hose reels).

4. *Smoke Clearance.* Provisions might need to be made to clear smoke and heat during as well as after fire-fighting operations to facilitate fire-fighting and rescue operations.

9.5.4.2 Other Considerations

9.5.4.2.1 *Capability/Capacity.* When considering the effect of fire department and emergency response force operations in a performance-based design, careful consideration should be given to the abilities of the local fire department.[20] These abilities can be divided into two categories—capability and capacity.

9.5.4.2.1.1 A fire department's *capability* is a measure of the department's ability to respond within a short time with sufficiently trained personnel and equipment to meet a set of objectives.

9.5.4.2.1.2 *Capacity* is a measure of a department's ability to respond adequately to multiple alarm incidents or simultaneous calls.

9.5.4.2.2 *Delay.* There can be significant delays associated with fire department response that should be considered when fire department operations are part of a performance-based design. These delays can come from the following sources[21]:

1. *Detection delay* is the delay associated with detecting a fire and initiating a response by the fire department.

2. *Dispatch delay* is the delay associated with receiving and processing an emergency call. The transition between detection delay and dispatch delay occurs when the fire department or emergency response force has direct, contractual control or approval authority of the equipment (e.g., telephone switching system and central station facility).

3. *Turnout delay* is the time it takes the responding emergency response personnel to react and prepare to leave the station. Turnout time stops when the wheels on the apparatus start to turn.

4. *Travel time* is the time at which the apparatus starts to move forward until the apparatus reaches the incident scene.

5. *Access time* is the time required for the department to move from the apparatus to the emergency location.

6. *Setup time* is the time required to prepare to commence operations. The level of effort (e.g., first responders or reinforcements) must be defined as part of the setup time. It is possible to have multiple setup times if fire response is staged.

7. *Suppression time* is the time required to suppress a fire.

8. Delays associated with apparatus already in service and awaiting call back for manpower are also considerations depending on staffing of the fire department.

9.5.4.2.3 *Impact of Fire Department Suppression Operations.* Fire departments can limit fire losses using two basic suppression tactics—extinguishment and containment. Extinguishment is commonly accomplished by direct application of water into the fire compartment. Figure 9–4 demonstrates a possible HRR for this

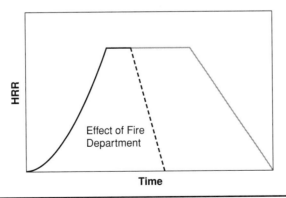

FIGURE 9–4 *Possible Effect of Fire Department Operations*

extinguishment tactic. Containment prevents fire extension to uninvolved portions of the building. Water is applied to surrounding walls and exposures while the involved contents are allowed to burn to completion.

Fire departments might not reach a building and start suppression operations until after flashover. In addition, suppression might be delayed because of higher priorities (e.g., rescue), or in some instances, suppression might not be the desired tactic. If the building is considered sacrificial and the contents of high value (e.g., artwork), the correct emergency response might be the removal of the contents rather than direct fire suppression. If this is the design approach, it might need to be documented in the performance-based design (PBD) report (see Section 12.2).

9.6   Occupant Behavior and Egress

9.6.1   *Objective.* To allow the occupants to travel safely to a place of safety in the event of a fire

9.6.2   **Concepts**

1. *Access.* Occupants should be able to access an exit.
2. *Protected Escape Route.* Exits might need to be fire rated to limit the potential for smoke and heat inflow until the last occupant can reach a place of safety.
3. *Protection in Place.* Protection can be provided so that occupants are safe inside the building.

9.6.3   For information on occupant behavior, see Section 8.5.3.

9.7   Passive Fire Protection

9.7.1   **Structural Stability**

9.7.1.1   *Objective.* To prevent the premature collapse of part or all of a structure

9.7.1.2   Concepts

1. *Inherent Stability.* Structural elements can be designed to withstand exposure to the expected fire severity without prematurely

collapsing. The design of the structural elements should take into account the loading on the structural element, the amount and duration of exposure, and the effects of deformation.

2. *Supplemental Protection.* If the inherent features of a structural element are not sufficient to maintain its strength, then protection can be applied directly to the structural element, or a barrier can be provided between the fire and the element to reduce the heat entering the structure.

Methods are available to determine the protection necessary for structural members to avoid structural collapse due to the thermal impact associated with exposure to elevated temperatures.[22,23,24] These calculation methods might consider either a standardized time–temperature relationship or a design fire scenario (see Chapter 8).

9.7.2   Limiting Fire Spread through Passive Means

9.7.2.1   *Objective.* To prevent a fire from spreading from the room of fire origin to other areas or compartments of the building

9.7.2.2   Concepts

1. *Compartmentation.* Buildings can be divided by fire barriers to isolate compartments that are not the room of fire origin from the effects of heat and smoke.

2. *Fire Barriers.* Fire barriers provide horizontal and vertical barriers to fire and smoke, typically provided in compartment boundaries. They are designed to resist the penetration of heat and smoke for a given amount of time.

3. *Protection of Openings.* Openings in barriers provide a ready avenue for fire spread. Fire spread through openings can be controlled by the following:

   • Firestop openings where piping, wiring, and conduit penetrate barriers.

   • Ensuring doors are closed in the event of a fire. This can be accomplished by using automatic closing mechanisms or educating occupants to close doors in the event of a fire.

   • Use of fire-resistant glazing materials. Ordinary glass has relatively little fire integrity.[25] Therefore, additional considerations should be made when glass is used in a barrier.

4. *External Spread.* Consideration should be given to a building's internal and external fire spread potential when a fire might breach the building envelope and attack upper floors.

5. *Controlling the Fire.* Automatic suppression systems can be used to assist in fire containment by reducing the severity of the fire.

References Cited

1. NFPA 550, *Guide to the Fire Safety Concepts Tree,* National Fire Protection Association, Quincy, MA, 1995.

2. Rockett, J. A., & Milke, J. A. "Conduction of Heat in Solids," *The SFPE Handbook of Fire Protection Engineering,* National Fire Protection Association, Quincy, MA, 1995.

3. Bukowski, R. "Fire Hazard Analysis," *Fire Protection Handbook,* 18th Ed., National Fire Protection Association, Quincy, MA, 1996.

4. Heskestad, G. "Fire Plumes," *The SFPE Handbook of Fire Protection Engineering,* 2nd Ed. National Fire Protection Association, Quincy, MA, 1995.

5. Zukoski, E., Kubota, T., & Cetegen, B. "Entrainment in Fire Plumes," *Fire Safety Journal,* vol. 3, pp. 107–121, 1980.

6. McCaffrey, B. "Purely Buoyant Diffusion Flames: Some Experimental Results," National Institute of Standards and Technology, NBSIR 79–1910, Gaithersburg, MD, 1979.

7. Thomas, P., Hinkley, P., Theobald, C., & Simms, D. "Investigations into the Flow of Hot Gasses in Roof Venting," Fire Research Technical Paper No. 7, HMSO, London, 1963.

8. Klote, J. H., & Milke, J. A. *Design of Smoke Management Systems,* American Society of Heating, Refrigerating, and Air-Conditioning Engineers, Atlanta, 1992.

9. NFPA 92A, *Recommended Practice for Smoke-Control Systems,* National Fire Protection Association, Quincy, MA, 1996.

10. NFPA 92B, *Guide for Smoke Management Systems in Malls, Atria, and Large Areas,* National Fire Protection Association, Quincy, MA, 1995.

11. Tewarson, A. "Generation of Heat and Chemical Compounds in Fires," *The SFPE Handbook of Fire Protection Engineering,* 2nd Ed., National Fire Protection Association, Quincy, MA, 1995.

12. Mulholland, G. "Smoke Production and Properties," *The SFPE Handbook of Fire Protection Engineering,* 2nd Ed., National Fire Protection Association, Quincy, MA, 1995.

13. Emmons, H. "Vent Flows," *The SFPE Handbook of Fire Protection Engineering,* 2nd Ed., National Fire Protection Association, Quincy, MA, 1995.

14. Schifiliti, R. P., Meacham, B. J., & Custer, R. L. P. "Design of Detection Systems," *The SFPE Handbook of Fire Protection Engineering,* National Fire Protection Association, Quincy, MA, 1995.

15. Evans, D., & Stroup, D. "Methods to Calculate the Response Time of Heat and Smoke Detectors Installed below Large Unobstructed Ceilings," NBSIR 85–3167, National Bureau of Standards, Gaithersburg, MD, 1985.

16. Schifiliti, R. P., Meacham, B. J., & Custer, R. L. P. "Design of Detection Systems," *The SFPE Handbook of Fire Protection Engineering,* 2nd Ed., National Fire Protection Association, Quincy, MA, 1995.

17. Fleming, R. "Automatic Sprinkler System Calculations," *The SFPE Handbook of Fire Protection Engineering,* 2nd Ed., National Fire Protection Association, Quincy, MA, 1995.

18. Puchovsky, M., Ed. *Automatic Sprinkler Systems Handbook,* 7th Ed. National Fire Protection Association, Quincy, MA, 1996.

19. AgentCalcs, Hughes Associates, Inc., Baltimore, MD, 1997.

20. Paulsgrove, R. "Fire Department Administration and Operations," *Fire Protection Handbook,* 18th Ed., National Fire Protection Association, Quincy, MA, 1996.

21. Barr, R. C., & Caputo, A. P. "Planning Fire Station Locations," *Fire Protection Handbook,* 18th Ed., National Fire Protection Association, Quincy, MA, 1996.

22. Milke, J. "Analytical Methods for Determining Fire Resistance of Steel Members," *The SFPE Handbook of Fire Protection Engineering,* 2nd Ed., National Fire Protection Association, Quincy, MA, 1995.

23. Fleishmann, C. "Analytical Methods for Determining Fire Resistance of Concrete Members," *The SFPE Handbook of Fire Protection Engineering,* 2nd Ed., National Fire Protection Association, Quincy, MA, 1995.

24. White, R. "Analytical Methods for Determining Fire Resistance of Timber Members," *The SFPE Handbook of Fire Protection Engineering,* 2nd Ed., National Fire Protection Association, Quincy, MA, 1995.

25. Fitzgerald, R. W. "Structural Integrity During Fire," *Fire Protection Handbook,* 18th Ed., National Fire Protection Association, Quincy, MA, 1996.

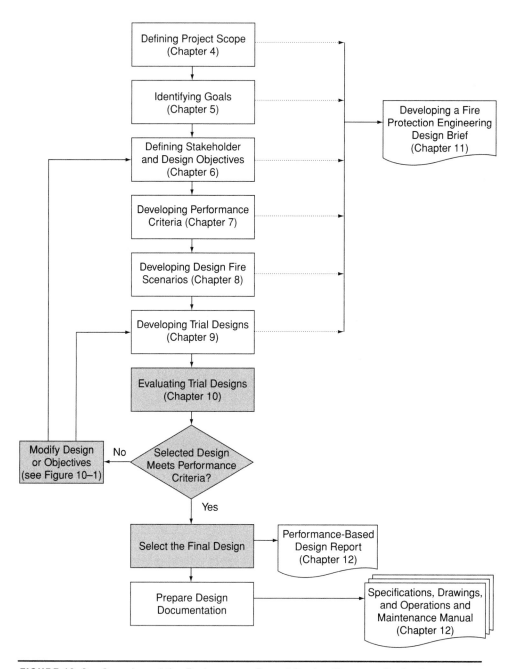

FIGURE 10–0 *Overview of the Performance-Based Design Process, Evaluating Trial Designs*

10

Evaluating Trial Designs

10.1 Overview of the Evaluation Process

10.1.1 Evaluation is the process of determining if a trial design (developed in Chapter 9) meets the performance criteria (developed in Chapter 7) when challenged by the postulated design fire scenarios (developed in Chapter 8).

10.1.2 The selected set of trial design(s) are tested against each design fire scenario. The intent is to demonstrate that in the design fire scenario, performance criteria will not be exceeded. If the trial design is successful, any remaining trial designs might be evaluated as necessary. If a trial design is not successful, the trial design might be modified and retested, or it might be dismissed. After the selected trial designs have been tested, a final design is selected from among those found successful. If there are no successful trial designs, the engineer should ensure that the trial designs considered all possible mitigation strategies. If after considering all possible mitigation strategies, there still are not any successful trial designs, the stakeholder objectives and performance criteria should be revisited (see Figure 10–1).

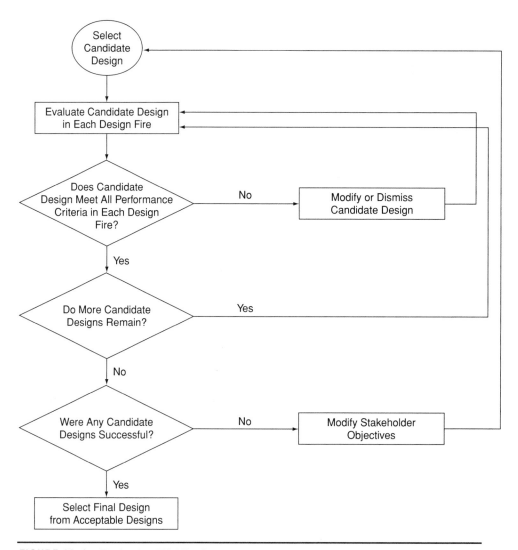

FIGURE 10–1   *Evaluating Trial Designs*

10.1.3   The technical detail needed varies with the evaluation level (see Section 10.2). System- or building-level applications might require independent evaluations of the individual fire protection components as well as a higher-order evaluation of the complete building design. Prior to beginning the evaluation process, the level of evaluation necessary should be established and agreed to by the stakeholders.

10.1.4   When considering whether a trial design meets specific performance criteria, factors such as effectiveness, reliability, availability, and cost should be considered.

10.1.4.1   The effectiveness of a trial design is judged by determining whether the design meets the established performance criteria.

10.1.4.2   *Reliability* measures whether a design or system will function as designed or intended. For example, a sprinkler system has functioned as designed if the system discharges sufficient water to control or extinguish a fire without excessive fire spread.

10.1.4.3   A design or system is considered *available* whenever it is capable of performing a required function at a given instant in time. Thus, the system might be unavailable during testing, unplanned maintenance, preventive maintenance, and planned modifications. A common failure for many fire protection systems is failure to restore to service following maintenance. Based on this definition of available, a system left inoperative following maintenance is considered unavailable. If this impaired system were required to operate and did not fulfill its design function, it would be an availability failure.

10.1.4.4   The combination of system reliability and availability is the system's *effectiveness.*

10.1.4.5   System costs can be used to evaluate trial designs. Costs that should be considered include initial installation costs, costs associated with inspection and maintenance, and costs associated with maximum acceptable fire damage. If the cost of a trial design is more than the client is willing to pay, consideration of it might not be necessary. However, if no trial designs fall within the client's budget, the stakeholder objectives should be reevaluated with the client. Also, if several successful trial designs are available, cost might be one of the factors used to select which successful alternative is used.

10.1.5   Time lines can be valuable tools for evaluating trial designs. Therefore, determination of the time of the following key events might be necessary:

- Ignition
- Fire detection
- Evacuation begins and ends
- Untenable conditions reached in room of origin
- Window failure
- Flashover
- Fire spread beyond room of origin
- Manual suppression
- Failure of structural elements
- Fire extinguishment

The key events will likely be related to the trial design option being evaluated and should be selected accordingly.

10.1.5.1   The heat release rate of the design fire curve is a key input into the calculation methods for determining the time of many of these events.

10.1.5.2   For some events, the activation of certain systems leads to a modification of the course of the design fire curve, which would otherwise eventuate. For example, if a growing fire activates sprinklers, then the heat release rate of the fire will be limited or reduced.

10.1.6   Many techniques can be used to evaluate the adequacy of a trial design. They fit into two principal categories—probabilistic and deterministic. A *deterministic analysis* examines the hazard posed by the potential design fire scenarios independently (see Section 10.4). A *probabilistic analysis* uses risk analysis to identify the consequences of specific events and their respective likelihoods. Two probabilistic methods are presented in this guide—the classical (explicit) risk analysis method and a risk binning analysis (see Sections 10.3.1 and 10.3.2, respectively). When completing an evaluation, there is an implicit assumption that an event more severe than that proposed is not credible or not within the scope of the PBD project.

10.2 Levels of Evaluation

10.2.1   The performance-based design process can be used to evaluate and recommend fire protection options at the subsystem-performance level, at the system-performance level, or at the building-performance level. Evaluations at each of these levels might be performed as comparative analyses or analyses and designs to defined performance criteria. The evaluation level necessary for a performance-based design project is a function of several factors. These factors include the following:

- Complexity of geometry
- Level of subsystem interaction
- Type of performance/acceptance criteria
- Sensitivity of subsystem output to design objectives
- Absolute or comparative evaluation
- Knowledge level (uncertainties)
- Benefit versus cost
- Expert judgment and experience

10.2.2 Subsystem Performance

10.2.2.1   A subsystem-performance evaluation typically consists of a simple comparative analysis and might be used to demonstrate that a selected component or subsystem provides *equivalent* performance to that specified by a prescriptive-based code. At this level, one sub-system is evaluated in isolation.

10.2.2.2   A comparative analysis typically uses analytical calculations to demonstrate equivalent performance of the proposed alternative component or system. Although these analyses might be deterministic or probabilistic, deterministic methods would normally be used. A deterministic, subsystem-performance evaluation might involve applying the same models, calculations, input data, and acceptance criteria for the trial design(s) and the code-mandated solution to which equivalence is sought.

For example, a subsystem-performance evaluation might be appropriate when considering an alternative fire detection system to that specified in a prescriptive-based code. Trial fire detection system designs might involve different types of detection devices, different spacing of detectors, or other changes to individual components or parts of the system. A trial design that resulted in detector activation and alarm signaling at a time or fire size equal to or earlier than the code-prescribed solution might be considered acceptable.

Similarly, a particular fire-rated structural element might be prescribed in a prescriptive-based code. A different structural element or a different approach to providing an equivalent level of fire resistance for a structural element might be considered acceptable in a performance-based analysis and design if the trial design can provide equivalent structural performance through a subsystem-performance evaluation.

10.2.2.3   In both of the previous examples, the evaluation was performed, and the results were compared to a prescriptive-based code. Evaluation of the detection system or the fire resistance in terms of the goals and objectives of the stakeholders rather than a prescriptive-based code is also possible. For a detection system, the objective might be to alert people in time for them safely to leave the building, or for fire resistance, the objective might be for the fire to be contained to the compartment of origin and the structure remain in place through a complete compartment burnout.

10.2.3 System-Performance Evaluations

10.2.3.1   A system-performance evaluation might consist of a comparative analysis or an analysis and design to performance criteria developed from goals and objectives of the stakeholders. A system-performance evaluation is used when an entire or a substantial part of a building design is being considered (e.g., the exit system) and more than one fire protection system or feature is involved (e.g., fire resistance ratings and suppression systems).

10.2.3.2   A system-performance evaluation is more complex than a subsystem-performance evaluation because the analysis needs to take account of the interaction between various systems and components. A system-performance analysis will often be based on one or more worst case scenarios (see Chapter 8).

10.2.3.3   In a system-performance comparative analysis, identical design fire scenarios should be used, and identical data, input parameters, and models must be used in both analyses. If this is done, any assumptions regarding rate of fire growth, choice of fire model, occupant characteristics, material response, or the like might not have a significant influence on the outcome. The analysis might be deterministic or risk based.

10.2.3.4   A system-performance evaluation might be appropriate for alternative design proposals that are not radically different from those included in prescriptive-based codes.

For example, a system-performance evaluation might be appropriate when a designer is proposing an atrium smoke management system with different performance criteria than specified in a code. In this situation, the engineer must analyze various fire and life safety issues, such as fire growth rates, smoke development, times for detection and suppression, smoke control system performance, and occupant egress. This analysis involves a number of fire protection systems and therefore requires a system-performance evaluation.

10.2.4   Building-Performance Evaluations

10.2.4.1   A building-performance evaluation is appropriate for complex or highly innovative buildings in which substantial analysis could lead to a clear understanding of the hazards and risks, to the resolution of complex design problems, and to the potential for significant cost savings. This building-performance analysis considers how the fire protection system interacts with the rest of the building. (An example might be using the potable water supply for the sprinkler system.)

10.2.4.2   Building-performance evaluations are typically probabilistic. By assigning probabilities of failure to the fire protection measures and assigning frequencies of occurrence to unwanted events, a probabilistic analysis can combine a number of different design fire scenarios as part of a complete fire safety assessment.

10.3 Probabilistic Analysis

10.3.1 Classical Risk Analysis

If the effect of probabilities and reliabilities is to be explicitly considered, a risk-based method must be used. A classical risk analysis is one technique that can accomplish this. This analysis requires a high degree of capability on the part of the engineers performing and reviewing the analysis. Also, because of the significant amount of statistical data and quantitative analysis, careful attention must be given to uncertainty.

10.3.1.1 Overview

In a classical risk analysis, the frequency of each design fire scenario and the reliability of each component of the fire protection system must be quantified. Possible sources of reliability and availability data include historical data for similar facilities, systems, or components (see Appendix E for methods to quantify risk).

10.3.1.2 Procedure

The steps required to complete a classical risk analysis are as follows:

- Develop design fire scenarios, and determine associated frequencies of occurrence.
- Determine the reliability of the trial design.
- Quantify the loss associated with the design fire scenario, assuming that the trial design is successful (see Section 10.3.1.3 for further discussion).
- Quantify the loss associated with the design fire scenario, assuming that the trial design fails (see Section 10.3.1.3 for further discussion).
- Repeat for each design fire scenario.
- Calculate the total risk associated with the trial design.
- Repeat for each trial design.

10.3.1.3 Success and Failure Modes

For some trial designs, there might be intermediate levels of success or failure. These intermediate modes can be an important strategy for fire risk reduction. Different failure modes of a single Chapter 9 subsystem (e.g., suppression) will result in different levels of loss. If the main valve of a sprinkler is improperly closed, the potential loss might be the entire building. If a sectional valve is improperly closed, the potential loss might be the associated section of the building. If a single sprinkler fails to actuate, a single room might be lost. When stated in terms of success, the outcomes might be as follows:

- A single sprinkler operation limits the damage to small area.
- A multiple sprinkler operation limits the damage to the room of origin.
- An extensive number of sprinklers activate and limit the fire to a single floor.
- An extensive number of sprinklers activate and do not control the fire, but they do provide early notification and delay fire spread until evacuation is complete.

The number of outcomes that must be considered is dependent on the project details and the required analysis accuracy.

10.3.2 Risk Binning Analysis

10.3.2.1 Overview

As an alternative to a classical risk analysis, the risk binning technique is simpler to apply. The importance of identifying all possible outcomes is less critical. By quantifying the consequences of the most severe events and coupling these with approximate event frequencies, an approximate, quantified risk estimate is possible.

10.3.2.2 Consequence Ranking

(a) A maximum consequence for each type of loss (e.g., occupant loss, monetary loss, business interruption, or environmental damage) should be identified. The consequences should represent the largest realistic event of each type.

(b) Each maximum consequence must be ranked. Table 10–1 provides an example of possible consequence-ranking thresholds (e.g., negligible, low, moderate, and high). The consequence predictions at this stage should bound (95 percent or better coverage) all possible event outcomes.

(c) The 95 percent coverage value is suggested because it has gained ready acceptance in other engineering fields.[1,2] By using this standard value, a comparison of different analyses is possible. If the use of an alternate coverage value is desired, all stakeholders should agree with this variation.

(d) When selecting the maximum consequence, an extensive analysis can often be avoided if the total replacement costs are assumed to be the maximum consequence.

10.3.2.3 Frequency Ranking

(a) The frequencies must also be ranked. Table 10–2 gives a sample description for frequency ranking.[3]

Table 10–1   Possible Consequence-Ranking Criteria

Consequence Level	Impact on Populace	Impact on Property/Operations
High (H)	Sudden fatalities, acute injuries, immediately life-threatening situations, permanent disabilities	Damage > $XX million Building destroyed, surrounding property damaged
Moderate (M)	Serious injuries, permanent disabilities, hospitalization required	$YY < damage < $XX million Major equipment destroyed, minor impact on surroundings
Low (L)	Minor injuries, no permanent disabilities, no hospitalization	Damage < $YY Reparable damage to building, significant operational downtime, no impact on surroundings
Negligible (N)	Negligible injuries	Minor repairs to building required, minimal operational downtim

Table 10–2   Example of Frequency Criteria Used for Probability Ranking

Acronym	Description	Frequency Level (Median Time to Event)	Description
A	Anticipated, expected	$>1 \times 10^{-2}$/yr (<100 yr)	Incidents that might occur several times during the lifetime of the building (incidents that commonly occur).
U	Unlikely	1×10^{-4}/yr $< f < 1 \times 10^{-2}$/yr (100–10,000 yr)	Events that are not anticipated to occur during the lifetime of the facility. Natural phenomena of this probability class include UBC-level earthquake, 100-year flood, and maximum wind gust.
EU	Extremely Unlikely	1×10^{-6}/yr $< f < 1 \times 10^{-4}$/yr (10,000–1,000,000 yr)	Events that will probably not occur during the life cycle of the building.
BEU	Beyond Extremely Unlikely	$<1 \times 10^{-6}$/yr (> 1,000,000 yr)	All other accidents.

(b) The frequencies should be for exceeding a specific loss (i.e., consequence) rather than for exceeding a specific scenario. Frequencies based solely on a specific scenario can be misleading. (A scenario might have a frequency of 1×10^{-7} per year. The conclusion is that *fire is not a concern*. However, the reported fire risk should represent the frequency of multiple fire scenarios. If 30 specific scenarios are developed, each at 1×10^{-7} fires per year, the net effect is 3×10^{-6} fires per year.)

(c) Alternate frequency rankings (bins) from those presented in Table 10–2 can be developed. Like the consequence rankings, all stakeholders should agree to the alternate values. It is even reasonable to add additional layers of ranking (e.g., single order of magnitude width rather than double orders of magnitude width).

10.3.2.4   Risk Ranking

(a) Once the bounding consequences and their respective frequencies have been estimated, they must be converted to a risk, which is accomplished by plotting the consequence–frequency combination on the matrix as shown in Figure 10–2. (The numbers in the boxes are for identification and do not imply a ranking.) Each consequence–frequency combination is assigned a relative risk level by the stakeholders. These resultant risks are considered bounding risks.

(b) After this analysis, events that meet certain risk criteria may be considered acceptable, depending on the stakeholders' objectives. For example, the stakeholders might consider moderate, low, and negligible risk events acceptable.

10.4 Deterministic Analysis

10.4.1 Prerequisites

10.4.1.1   In a deterministic analysis, the expected performance of the fire protection system is analyzed against one or more design fire scenarios. When multiple design fire scenarios have been developed, they should be considered independently.

10.4.1.2   Time lines can be useful in a deterministic analysis. Descriptive factors of the design fire scenario (e.g., ignition, growth, and critical heat release rates) can be plotted on the time line along with the times of occurrence of key events of the fire protection system.

Frequency→ Consequence↓	Beyond Extremely Unlikely $f \leq 10^{-6}\,\text{yr}^{-1}$	Extremely Unlikely $10^{-4} \geq f > 10^{-6}\,\text{yr}^{-1}$	Unlikely $10^{-2} \geq f > 10^{-4}\,\text{yr}^{-1}$	Anticipated $f > 10^{-2}\,\text{yr}^{-1}$
High		7	4	1
Moderate	10	8	5	2
Low		9	6	3
Negligible	11	12		

Key

■ High risk   ▨ Moderate risk   ☐ Low risk   ☐ Negligible risk

FIGURE 10–2   *Example of a Risk-Ranking Matrix*

10.4.1.3   For the trial design to be successful, each performance criterion must be met in each of the design fires with consideration given to uncertainties due to known variations and unknown effects. Factors that might introduce uncertainty into the analysis include material variations, installation unknowns, system and component variability, unanticipated use of systems, and unpredictable future human actions.

10.4.2   Fire models or other types of analytical analysis are often used as the basis for deterministic analysis. Appendix F provides information on the selection of models or other analytical methods. The analytical methods used should be capable of determining if the performance criteria will be achieved in the design fire scenarios. The following diagrams identify possible analysis procedures for determining if different types of performance criteria have been met.

10.4.2.1   Prevention of Fire Spread

Figure 10–3 provides a possible methodology for considering fire spread from enclosures.

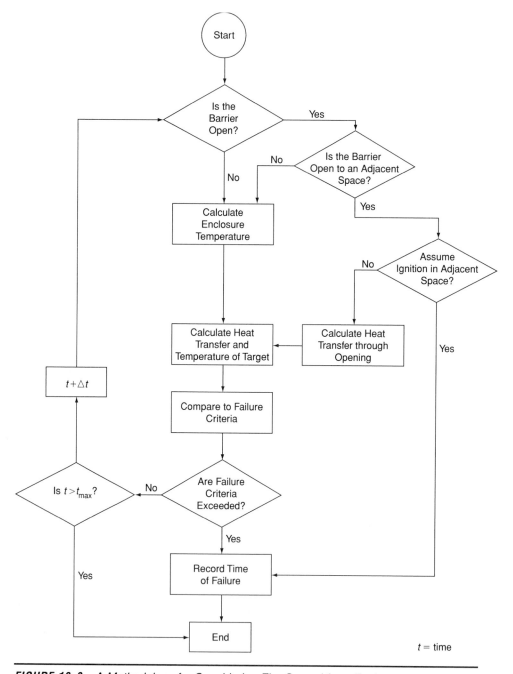

FIGURE 10-3   *A Methodology for Considering Fire Spread from Enclosures*

10.4.2.2   Fire Detection

Figure 10–4 provides an overview of the process of estimating the detector activation time for a given design fire scenario. When using this methodology, note that in some cases a design fire scenario might not activate a detector. The methodology presented below is essentially the same process as the computer model DETACT[4] uses.

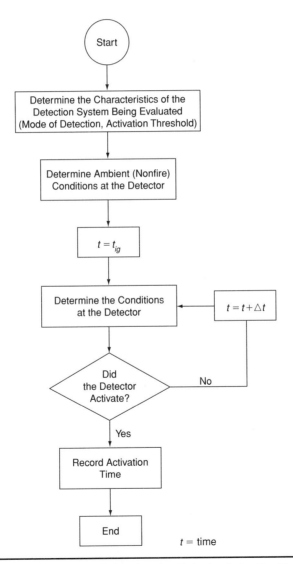

FIGURE 10–4   *A Methodology for Estimating the Detector Activation Time*

10.4.2.3   Occupant Evacuation

Figure 10–5 provides an overview of the process for estimating evacuation time.

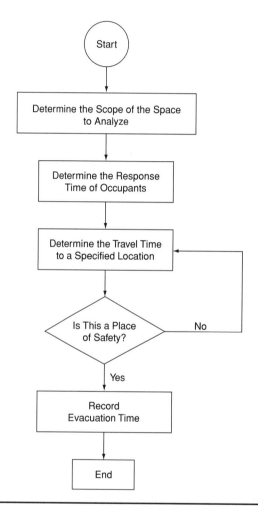

FIGURE 10–5   *A Methodology for Estimating Evacuation Time*

10.4.2.4 Smoke Development Performance Criteria

Figure 10–6 provides an overview of the process for evaluating the achievement of smoke development performance criteria.

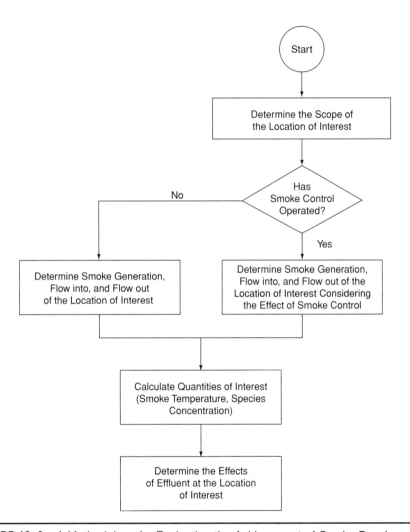

FIGURE 10–6 *A Methodology for Evaluating the Achievement of Smoke Development Performance Criteria*

10.5 Accounting for Known Variations and Unknown Effects

10.5.1 Overview

10.5.1.1   The engineering design process must provide a level of assurance that the designed system or component achieves its performance criteria. The analysis must account for expected variations in materials, components, demands, performance, and occupant characteristics. In prescriptive-based code designs, an unstated excess capacity is used to address variations. (The required design might actually perform better than necessary.) Prescriptive-based code designs with an engineering basis might prescribe a minimum safety factor that accounts for unknowns. This safety factor is derived from either practical experience or estimates of appropriate conservatism. The safety factor commonly applied in a structural stress analysis is an example.

10.5.1.2   For performance-based designs, an allowance must be included to account for unknowns and variations and to provide a level of confidence in the final design. This allowance might be based on historically derived safety factors, an uncertainty analysis, or analysis margins. Each of these techniques is described below.

10.5.1.3   Assumptions, limitations in calculation procedures, and variability in actual versus evaluated scenarios will result in uncertainty. Therefore, including explicit uncertainty or safety factors in deterministic-based system-performance evaluations might be appropriate. However, avoid unnecessary or excessive use of safety factors if the basic assumptions and calculation procedures are known or have been demonstrated to be highly conservative.

10.5.1.4   Details on available computer models might not allow explicit evaluation of uncertainty. Conservative approaches to design fire scenarios and performance criteria might be utilized to address uncertainty and safety factors.

10.5.1.5   The engineer, in consultation with other stakeholders, should determine whether it is appropriate to include explicit safety factors within the evaluation, or whether the assumptions and calculation procedures are intrinsically and sufficiently conservative.

10.5.1.6   When safety factors are not used, the engineer should understand not only the theory behind uncertainty or margin analysis but also its application to a complex fire protection engineering design.

10.5.2 Uncertainty in the Engineering Design Process

Performance-based fire protection analysis and design relies on current scientific knowledge and the ability to perform accurate technical predictions. Predictions of the buildup of temperature and combustion products in a building as well as the time required to evacuate the building safely are often made. Therefore, the consideration of several key sources of uncertainty in the design process and resulting predictions is important because uncertainty can take many forms. These key sources are as follows:

- Uncertainties about the science and engineering being used
- Uncertainties about human behaviors during a fire
- Uncertainties about risk perceptions, attitudes, and values

For instance, there can be uncertainty regarding the following issues:

- The physical parameters such as the ignition source, the flame spread rate, and the heat release rate of furnishings
- The appropriateness of a particular tool for a particular prediction
- Human behavioral response to a fire
- The age or health of people who might be exposed to a fire
- Individual and societal attitudes and values

10.5.2.1 Uncertainties in Science and Engineering[5]

(a) *Theory and Model Uncertainties.* In engineering correlations and fire models, theory uncertainty relates to the accuracy and appropriateness of an equation or correlation to the problem being addressed. Due to the limits in our scientific knowledge and computing power, all correlations and calculation procedures make simplifying assumptions. Engineering model predictions are based on correlations (e.g., curve fits to empirical data) and scientific-based calculations (both manual and computer-based). Both correlations and calculations are approximations of reality. Deviations between reality and the model predictions are considered model and theory uncertainty. The appropriate questions regarding theory and model uncertainty relate to the similarity of the design conditions compared to those under which the correlation data were taken, and how they relate to the appropriateness of assumptions and limitations inherent in scientific calculations compared to those of the design.

(b) *Data and Model Inputs.* Much of the input data relating to material properties, fire growth curves, and species production rates are subject to uncertainty. Using a range of values and not simply a single input data point is often necessary. Determining when a range of values is necessary and how to select appropriate input values is discussed in Appendix G.

(c) *Calculation Limitations*. Not only are the assumptions inherent in the calculations important to the appropriateness of the use of a tool, but the boundary conditions used in modeling also have considerable potential impact. For example, for most fire models, different answers result depending on the control volume selected for modeling and the level of detail of the model. Also significant to the uncertainty in the model predictions are the index variables used to identify a location in the domain of a model or to make calculations specific to a population or geographic region.

(d) *Design Fire Scenario Selection*. Differences between reality and the design fire scenarios, which were used to evaluate a trial design might exist. The representativeness of the design fire scenarios and their respective design fires should be evaluated.

10.5.2.2   Uncertainties in Human Behaviors

Uncertainty is associated with the human element in performance-based design, from the perspective of how people react in fires to the ability to predict how future human actions might impact the design conditions.[5]

10.5.2.3   Uncertainties in Risk Perceptions, Attitudes, and Values

Uncertainties in risk perceptions, attitudes, and values involve determining what is important to the stakeholders and to what lengths they will go to provide protection. Because most projects have many stakeholders such as the building owner, the engineer, the architect, the code official, and the public (users of the building), assigning worth to the usefulness or importance of something and applying it to both individual and societal issues is difficult. Uncertainties about risk perceptions, attitudes, and values are often at the center of discussions about how safe is safe enough. Decisions that change if a value, attitude, or risk perception varies must be made explicit in the design.[5] Agreement on these key decisions by all stakeholders is critical to the success of a PBD. Chapters 5, 6, and 7 describe how to make values and attitudes explicit in the predictions.

10.5.3 Safety Factors

10.5.3.1   Safety factors are normally historically derived. If an adequate historical data set exists (i.e., a large number of systems that have performed successfully in mitigating a fire), the overcapacity of each successful system can be estimated. The minimum overcapacity might be the basis for the selection of the safety factor.

10.5.3.2   The use of safety factors to achieve the required excess capacity is best suited for deterministic analyses. For most situations, the deterministic analysis should be completed using nominal values, when they are the mean values from experimental results. For some situations, the use of other values (e.g., the maximum number of occupants) might be necessary. The types of values used will be dependent on how the safety factor was derived.

10.5.4   Classical Uncertainty Analysis[2]

(a) Uncertainty analysis techniques were developed to estimate the expected error and variations that occur during measurement and tests. The primary purposes of these analyses is to predict the pre-experiment expectations (e.g., whether the experiment will provide useful results), estimate the accuracy of an instrumentation system (e.g., whether the instrument will provide acceptable information), and estimate the uncertainty of a resultant (e.g., the confidence interval for a test result).

(b) In measurement and testing, uncertainty is classified as random or systematic error.

(c) Appendix G describes calculation methods that might be used in a classical uncertainty analysis.

10.5.4.1   The uncertainty prediction can be the basis for the design allowance. If the reliability must be 95 percent, then the predicted value must include an allowance equal to or greater than the uncertainty. If the reliability must be 99 percent, then the predicted value must include an allowance equal to or greater than 1.5 times the uncertainty.[2]

10.5.4.2   It is possible to simplify an uncertainty analysis by determining which variables contribute the most significantly to the overall uncertainty. If some variables have a low effect on the results of the analysis, the effect of these variables on the uncertainty might be neglected.

10.5.4.3   The uncertainties in a performance-based design can be grouped as input uncertainties, construction and use variations, calculation methodology limitations, and unrecognized behaviors.

10.5.4.3.1   *Input Uncertainties*

(a) Ranges of certainty can be established for material properties, failure thresholds (e.g., flashover temperature criteria), and equipment performance expectations.

(b) For example, a pump will be expected to flow a specific volume of water at a specific pressure (0.20 ± 0.02 m^3/s for a discharge pressure of 800 kPa). Thus, the expected flow might range from 0.18–0.22 m^3/s. Typically, these ranges are reported at 95 percent confidence. ASME PTC 19.1, *Test Uncertainty*,[2] provides additional details on reporting uncertainties that might be considered inputs to a performance-based design.

(c) Other examples of input data uncertainty include material properties, fire growth curves, and species production rates.

(d) When using classical uncertainty analysis, the random uncertainties for widely known and developed parameters (e.g., acceleration of gravity) should be based on the mean standard deviation, $S_{\bar{x}}$. For less well-established values, the random uncertainties should be based on the standard deviation of the known data set or sets.

10.5.4.3.2 *Construction and Use Variations*

(a) Ranges can be established for construction (e.g., construction tolerances), material variations (e.g., accepted variations in performance, dimensions, and density), and operating conditions (e.g., combustible loading and occupant loading).

(b) When using classical uncertainty analysis, the random uncertainties for construction and use variations should be based on the standard deviation of the known data set or sets.

10.5.4.4 An alternative to the classical uncertainty analysis is presented in Appendix G.

10.5.5 Analysis Margin

10.5.5.1 The confidence that an analyst has in his or her results will vary with many factors. The confidence can be referred to as the *analysis margin*. The higher the margin, the more confidence that the prediction will never be exceeded. (Higher margins also equate to more expensive designs.) Qualitative descriptions of confidence include bounding, sufficiently bounding, and best estimate. Each of these terms will be defined below.

(a) *Best Estimate.* When nominal or average data are used for the analysis, the results are described as best estimate. To assume that the predictions will be low 50 percent of the time and high 50 percent of the time is usually acceptable.

(b) *Bounding.* When one or more parameters used for the analysis are set to such an extreme that the results are biased to a conservative result (e.g., if a thermally based performance criterion is specified, the use of the highest observed heat release rate as the energy release of the design fire), the results are described as bounding. The selected parameter set at the extreme must provide the most conservative result relative to the performance criteria.

(c) *Sufficiently Bounding.* A result is considered sufficiently bounding when all but one parameter used for an analysis are set to best-estimate values, and the one extreme parameter is set as follows:

- Values associated with operating expectations (e.g., combustible loading and number of building occupants) should be taken at a minimum of 90 percent of the anticipated situations (i.e., coverage). For example, for fixed stadium seating, the 99, 95, and 90 percent coverages might be considered equal to the number of seats if sellouts are anticipated at a majority of the events.
- Scientific input values (e.g., peak heat release rate and flashover temperature) should be taken at 95 percent coverage.

10.5.5.2   Sufficiently bounding or bounding results should be used in deterministic and risk binning consequence predictions. Best-estimate results should be used in risk binning frequency and explicit risk method predictions.

10.5.6 Other Evaluation Techniques

Several analysis techniques that might be useful in an evaluation are available. Several of the more useful are summarized below.

10.5.6.1   Importance Analysis[6]

A process by which each analysis parameter is assigned a numerical ranking on a relative scale from zero to one. An importance value of zero indicates that the variable has no effect on the uncertainty results. A value of one implies total correlation when all of the uncertainty in the results is due to the uncertainty of a single parameter. This process is especially useful for demonstrating that the inaccuracy of specific variables does not produce a significant inaccuracy in the results.

10.5.6.2   Sensitivity Analysis[7]

This process determines how changes in one or more parameters of an analysis change the results and conclusions (i.e., the resultant). This is accomplished by holding all parameters constant while varying a single parameter. The effect of this variation on the output can then be reported. By testing the responsiveness of the results to variations in the values assigned to different input parameters, sensitivity analysis allows the identification of those input parameters having the most significant effect on the predicted resultant. A sensitivity analysis does not tell the decision maker the value that should be used, but it shows the impact of using different values. A sensitivity analysis will perform the following functions:

- Identify the dominant variables in an analysis
- Demonstrate the sensitivity of the results to variation in input data

10.5.6.3   Switchover Analysis

This is a process in which one or more inputs is iteratively varied in order to find the values (if any) of the inputs that would cause a strong enough change in the value of the resultant to change the final decision.

10.5.6.4   Parametric Analysis

In parametric analysis, detailed information is obtained about the effect of a particular input on the value of the outcome criterion.

10.5.6.5   Comparative Analysis

This technique evaluates risks and costs in order to mitigate risk by means of comparison to other similar risks.

10.5.6.6   Expert Elicitation

If hard data do not exist and cannot be created, often an expert elicitation is conducted in order to obtain expert judgment of an uncertain quantity.

References Cited

1. American National Standards Institute, *U.S. Guide to the Expression of Uncertainty in Measurement,* 1997, National Conference of Standards Laboratories, Boulder, CO.

2. ASME PTC 19.1, *Test Uncertainty: Instruments and Apparatus,* American Society of Mechanical Engineers, New York, 1998.

3. *Preparation Guide for U.S. Department of Energy Nonreactor Nuclear Facility Safety Analysis Reports,* U.S. Department of Energy, Washington, DC, July 1994.

4. Evans, D., & Stroup, D. "Methods to Calculate the Response of Heat and Smoke Detectors Installed Below Large Unobstructed Ceilings," National Institute of Standards and Technology, Gaithersburg, MD, 1985.

5. Morgan, M., & Henrion, M. *Uncertainty: A Guide to Dealing with Uncertainty in Quantitative Risk and Policy Analysis,* Cambridge University Press, Cambridge, NY, 1998.

6. Phillips, W. *Tools for Making Acute Risk Decisions with Chemical Process Applications,* Center for Process Safety, New York, 1995.

7. Phillips, W. "Computer Simulation for Fire Protection Engineering," *The SFPE Handbook of Fire Protection Engineering,* 2nd Ed., National Fire Protection Association, Quincy, MA, 1995.

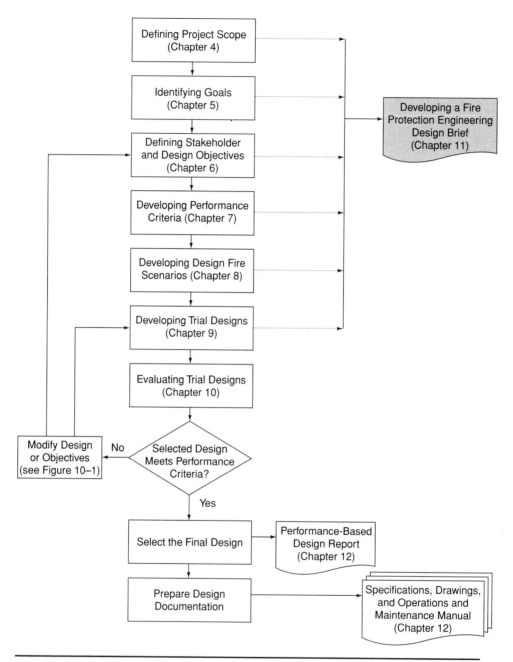

FIGURE 11–0 *Overview of the Performance-Based Design Process, Developing a Fire Protection Engineering Design Brief*

11

Developing a Fire Protection Engineering Design Brief

11.1 General

11.1.1 The objective of the Fire Protection Engineering Design Brief is to review the architectural proposals, identify potential fire hazards, and define the fire safety problems in qualitative terms suitable for detailed analysis and quantification.

11.1.2 At the beginning of the performance-based analysis and design process, the engineer and other stakeholders should do the following:

- Agree on the project scope, design intent, and building use
- Define and prioritize fire safety objectives and performance criteria
- Agree on how the fire safety goals and objectives, as they pertain to the design intent, should be evaluated in accordance with this guide
- Determine the implementation of the selected solutions into the project scope and execution
- Establish the necessary verification and quality assurance to ensure compliance with the agreed to solutions

The resulting documentation of this process is called the *Fire Protection Engineering Design Brief.*

11.1.3   The Fire Protection Engineering Design Brief might be developed by the engineer and presented to other stakeholders for approval or developed collaboratively between the engineer and one or all of the other stakeholders.

11.1.4   By reaching an agreement on the design approach prior to analysis, the engineer ensures that effort is not expended on any designs or evaluation methods that will not be acceptable to one or more stakeholders.

11.1.5   The Fire Protection Engineering Design Brief will usually be created after the architectural concept design has been completed. The Fire Protection Engineering Design Brief is created after the engineer has reviewed the project, had initial contact with other stakeholders, and planned a course of action. The Fire Protection Engineering Design Brief documents the agreed upon goals and objectives to the point of concerted analytical analysis and creates a statement of understanding among all the parties involved.

11.1.6   Purpose of the Fire Protection Engineering Design Brief

11.1.6.1   The purpose of the Fire Protection Engineering Design Brief is to document and assist in reaching agreement between the engineer and all stakeholders on the preanalysis portions of the design prior to commencing the quantitative design and analysis.

11.1.6.2   The Fire Protection Engineering Design Brief documents that all parties involved agree on the performance criteria and the methods that will be employed to evaluate the trial designs. The creation of the Fire Protection Engineering Design Brief will improve communications among stakeholders early in the design in order to prevent any misunderstandings at a point when the resolution might not be easily achieved.

11.2 Contents of the Fire Protection Engineering Design Brief

11.2.1   The Fire Protection Engineering Design Brief includes the following items:

11.2.1.1   Definition of the project scope (see Chapter 4)

11.2.1.2   Description of the building and occupant characteristics (see Sections 8.2.3.3 and 8.2.3.4)

11.2.1.3   The project goals (see Chapter 5)

11.2.1.4   The project objectives (see Chapter 6)

11.2.1.5   The performance criteria (see Chapter 7)

11.2.1.6   The selected fire scenarios (see Section 8.1–8.4)

11.2.1.7   One or more trial designs (see Chapter 9)

11.2.1.8   The levels and methods of evaluation (see Section 10.2)

11.2.1.9   A record of agreement on the above

11.2.2   The materials included in the Fire Protection Engineering Design Brief are not necessarily formal submittals, but they are ultimately included in the documentation of the performance-based design. The following materials should be included.

11.2.2.1   Documentation of Project Participants

11.2.2.1.1   A list identifying all parties participating in the project should be created. Identifying the appropriate representatives of the stakeholders that have the authority and ability to prioritize fire safety goals and recording the names and positions of the individuals is important. Stakeholders that should be documented include, but are not limited to, the following:

- Client
- Engineers
- Architects
- AHJs
- Insurance companies
- Contractors

11.2.2.1.2   The list of participants should include their business addresses, mailing addresses, phone numbers, fax numbers, the particular role each person will play in relation to the decision-making process, and his or her authority.

11.2.2.2   Documentation of Qualifications

11.2.2.2.1   The engineer's qualifications should be documented. The form of this documentation will depend on past working relations with the stakeholders and the requirements of state and local law. When presenting the idea of a performance design for the first time to a stakeholder, a resume, which discusses relevant experience and lists a number of projects of similar scope and magnitude, might need to be provided. This presentation of qualifications might be the only method the stakeholder has to determine if he or she will allow the engineer the latitude to prepare a performance design.

11.2.2.2.2 Professional registration such as Professional Engineer (i.e., P.E. or P.Eng.) or Chartered Engineer (i.e., Ch.E.) might be the only requirement to conduct engineering work in a given jurisdiction. However, more qualifications might be necessary to persuade the stakeholder that the engineer is qualified to perform a performance-based design. These additional qualifications might include educational experience, illustration of work on similar projects, and membership in professional societies or technical committees.

11.2.2.3 General Project Information

11.2.2.3.1 After the engineer has demonstrated his or her qualifications for undertaking the performance design, the engineer must agree with other stakeholders on a range of assumptions and methods necessary to perform the calculations for a performance design. These issues include the following:

- *Project Scope.* The project scope clearly defines the borders of the performance design. The scope might include a part of a building, an entire building, or multiple buildings. This will document the engineer's area of responsibility within the overall project.

- *Purpose of Design.* The purpose might be to ensure that the design meets the stakeholder objectives or to evaluate an alternative to a code-specific requirement. For example, each of the three model building codes has a section that allows alternate materials or methods to be used in the design of a building. This section might be referenced to show the purpose of the performance-based design.

- *Identification of Objectives and Performance Criteria.* The stakeholder objectives, the design objectives, and the performance criteria must be identified. The critical items that will be calculated and presented as part of the performance design must be reviewed with the appropriate stakeholder to gain concurrence on their inclusion in the report.

11.2.2.4 The level of sophistication and the methods used to document the performance design will vary depending on the relationship the designer has with other stakeholders and the scope of the project. The method used to document the design might depend on what the stakeholders want. Regardless of the method used, all the decisions, formal and informal, should be documented in the final report. The following items are examples of this documentation:

- Minutes of meetings and telephone conversations that summarize what factors have been agreed upon

- A formal letter asking for permission to conduct the evaluation in a certain manner, signed and returned or answered formally by stakeholders

- A notation in a log book describing a telephone conversation and a simple understanding of what will be done

11.2.2.5   The Fire Protection Engineering Design Brief is a dynamic document that can be updated as additional information becomes available (e.g., increased fire loads, project scope changes, and changes allowing nonambulatory occupants where only ambulatory occupants were planned). The Fire Protection Engineering Design Brief may become the basis for initial chapters of the performance-based design report.

11.3 Submittals

11.3.1 Submittals for approval might be made to the stakeholders following each step indicated in Section 12.2, following the completion of several steps, or following the completion of the Fire Protection Engineering Design Brief.

11.3.2 A submittal schedule can be developed by the engineer to document which stakeholders review which portions of the Fire Protection Engineering Design Brief. The submittal schedule is a tool that follows the progress of the design and obtains any necessary approvals at critical milestones.

11.3.3 The submittal schedule should list each project deliverable and the person(s) who must approve the document. Multiple approvals might be necessary. Not every stakeholder must approve each document. For some projects, the approval order might be of concern. If this is the case, then this information should be part of the submittal schedule. The submittal schedule should also specify who would get a copy of the final documentation, intermediate deliverables, and the time they are expected.

11.4 Following completion and approval of the Fire Protection Engineering Design Brief, the engineer will begin the evaluation portion of the design. This will include the development of the complete characteristics of the design, quantification of the design fire(s), and evaluation.

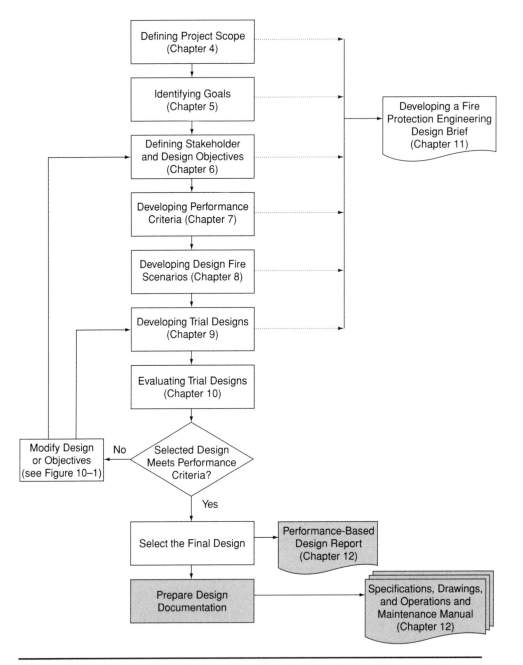

FIGURE 12–0   *Overview of the Performance-Based Design Process, Documentation and Specifications*

12

Documentation and Specifications

12.1 Introduction

12.1.1 Proper documentation of a performance design is critical to the design acceptance and construction. Proper documentation will also assure that all parties involved understand what is necessary for the design implementation, maintenance, and continuity of the fire protection design. If attention to details is maintained in the documentation, then little dispute during approval, construction, start-up, and use should occur.

12.1.2 Poor documentation could result in the rejection of an otherwise good design, poor implementation of the design, inadequate system maintenance and reliability, and an incomplete record for future changes or if the design were forensically tested.

12.1.3 The documentation has four parts—the Fire Protection Engineering Design Brief (see Chapter 11), the performance design report, the detailed specifications and drawings, and the building operations and maintenance manual. Although each part has its own unique requirements, there might be overlapping documentation requirements. Also, each part might be combined with other parts. The documentation of each project will largely be determined by the requirements of the stakeholders and the unique aspects of the project.

12.1.4   The documentation should be brief and concise. A clear writing style ensures that all important information is not lost in the text or misinterpreted.

12.1.4.1   Also, all critical design information must be included. For example, if the design is based on a critical fire load not being exceeded, that fire load must be maintained throughout the life of the building. If the design is predicated on the proper operation of a suppression system and its reliability, then the maintenance of the system and the proper ways to handle a service interruption must also be included.

12.2   Performance-Based Design Report

12.2.1   Due to the importance of the design report, it must be a thorough, clear, and unambiguous document. Many stakeholders might review this report, including the AHJ and his or her staff, an appeals board, the building owner, the building insurer, the building operator, a future building purchaser, or a forensic team after a fire in the building has occurred. Because some of these reviewers might have limited technical or fire protection training, the design report should be prepared for a general audience. The report should convey the expected hazards, risks, and system performance over the entire building life cycle (i.e., construction, operation, renovation, and demolition).

12.2.2   Generally, the report might, but need not necessarily, follow the outline below. The particular presentation of this information may vary based on the author's personal editorial style. It should include the following items:
- *Project Scope.* The project scope defines the extent of the project (e.g., part of a building or multiple buildings). General information and assumptions relative to the design should be included, such as building characteristics, occupant characteristics, and any existing measures in place.

- *Engineer's Capabilities.* A resume and other information supporting the qualifications of the engineer(s) performing the analysis should be provided.

- *Goals and Objectives.* The fire safety goals and objectives agreed between the engineer and other stakeholders should be included in this section. How the stakeholder objectives were developed should be discussed. Also, how the design objectives were developed, including any uncertainty and safety factors, should be included.

- *Performance Criteria.* Each performance criterion should be included in this section. How the performance criteria were developed, including any uncertainty or safety factors, should be discussed.

- *Fire Scenarios and Design Fire Scenarios.* Each fire scenario should be discussed in this section. The selection of design fire scenarios and the basis used to select those fires should be discussed. The discussion should include the expected conditions under which the design will be valid.

- *Final Design.* The final fire protection systems and controls that were selected from the alternatives and will be used in the construction of the building should be described with a discussion of how they meet the performance criteria.

- *Evaluation.* The evaluation of the final design should be thoroughly discussed. This discussion should include a description of the evaluation method, the design tools used, and the establishment of uncertainty and safety factors. Any computational tools should be discussed along with the input and output. For some projects, a description of the trial designs that were not selected and the reasons for their rejection might be appropriate.

- *Critical Design Assumptions.* This section includes all the assumptions that must be maintained throughout the life cycle of the building in order for the design to function as intended.

- *Critical Design Features.* This section includes all the design parameters that must be maintained throughout the life cycle of the building in order for the design to function as intended.

- *References.* Selected references should be part of the design documentation, especially those that are proprietary or difficult to obtain. Also, those that need to be available to support O&M should be provided. Stakeholder agreement on this selection is important.

12.3 Specifications and Drawings

12.3.1 The specifications and drawings convey to building and system designers and installing contractors how to implement the performance design. Specifications and drawings might include required sprinkler densities, hydraulic characteristics and spacing requirements, the fire detection and alarm system components and programming, special construction requirements including means of egress and location of fire-resistive walls, compartmentation, and the coordination of interactive systems. The detailed specifications are the implementation document of the performance design report. The specifications are derived from the calculations and results within the report. The transformation of the concept design into a construction document is of critical importance. A mistake here, such as the transposition of numbers, the dropping of decimal places, or any other inadvertent mistake, could have consequences if the performance design is tested under actual fire conditions.

12.3.2 Form of Specifications

12.3.2.1 Attention to detail must be maintained. Master specifications or guide specifications might not be suitable due to the unique features of the performance design. Simply referencing existing codes might not be adequate to assure the quality necessary to implement the fire protection performance design. However, portions of prescriptive-based codes and guide specifications might apply. If the fire protection design is predicated on a particular spacing of sprinklers or sprinkler design density, then the characteristics and spacing requirements of that sprinkler system must be clearly indicated.

12.3.2.2 The specifications will establish the minimum acceptable characteristics of the fire protection systems. Thus, they provide the basis for proper execution of the report recommendations.

12.3.2.3 The specifications should establish the required quality controls necessary to ensure that the fire protection systems will achieve their intended function.

12.3.2.4 Commissioning of fire protection systems and review of their installation to validate that the installed fire protection systems meet the proposed intent of the design is essential to the level of fire safety provided in the structure. The engineer should be involved in the production and review of design documents, review of shop drawings, field inspections, and acceptance testing of the fire protection systems.

12.4 Detailed Drawings

The detailed drawings will graphically represent the results of the performance design. Detailed drawings might include required exit widths, construction features of fire-resistive walls, the location of fire protection devices, and the design of fire protection systems. The detailed drawings create a visual record of the performance design.

12.5 Operations and Maintenance Manual

12.5.1 The fire protection operations and maintenance manual clearly states the requirement of the building operator to ensure that the components of the performance design are in place and operating properly. The O&M manual also describes the commissioning requirements and the interaction of the different systems interfaces. All subsystems are identified, and inspection and testing regimes and schedules are created. Of primary concern is the documentation of the proper function of subsystem interactions. Although some systems might be tested and inspected individually, the interconnections between systems should be periodically tested. This includes elevator recall, air handler shutdowns and smoke control, and locking door release.

12.5.2 The O&M manual also gives instructions to the building operator on restrictions placed on building operations. These limitations are based on the engineer's assumptions during design and analysis. These limiting factors might include critical fire load, sprinkler design requirements, building use and occupancy, and reliability and maintenance of systems. The design components that are critical to the achievement of the goals must be maintained, and a maintenance plan for those components must be developed.

12.5.3 Part of the O&M manual will be used to communicate to tenants and occupants these limits and their responsibilities as a tenant. It might also be used as a guide for tenant renovations and changes.

12.5.4 The O&M manual can be used to document agreements with the stakeholders. These agreements might include inspection frequencies, inspector qualifications, a fee structure for unique AHJ inspections, or a method for selecting a third party inspection service.

12.5.5   The O&M manual should include a statement highlighting the fact that actions might need to be taken if a fire protection system is impaired or removed from service. The selection of compensatory measures might depend on the magnitude and duration of the impairment, such as posting a fire watch during fire alarm system maintenance or reducing the combustible loading if a sprinkler system is removed from service for an extended period of time. The compensatory actions form an integral part of the overall performance-based design and therefore must be clearly understood and agreed to by all stakeholders. Proper documentation of the compensatory actions is essential so that the building operators can readily implement them in the event of a system impairment.

12.6   Testing Documentation

12.6.1   The engineer should develop testing criteria for the stakeholders' review. Included in the testing criteria should be a method that documents the procedure and pass/fail criteria. Testing criteria should include acceptance testing and periodic testing to assure system performance.

12.6.2   Documentation of testing and its results should be maintained with the building records.

12.7   Proper documentation will increase the likelihood that the design will be accepted by the stakeholders and will assure that the design is properly implemented and maintained for the life cycle of the building. The documentation must be presented in a clear manner that accommodates all audiences. Although assumptions might be self-evident, they must be documented to assure agreement by all participants. These documents must be available to all parties throughout the life of the building to assure continuity of the fire protection features included in the design.

Appendix **A**

Selected Readings

Performance-Based Fire Protection Engineering

Fire Engineering Guidelines, Fire Code Reform Centre Limited, Sydney, NSW, Australia, 1996.

"Fire Safety Engineering in Buildings," DD 240, British Standards Institute, London, 1997.

ISO CD13887, *Fire Safety Engineering,* International Standards Organization, Geneva, 1997.

Reiss, M. "Global Performance-Based Design: Is It the Solution?" *Proceedings, 1998 Pacific Rim Conference and Second International Conference on Performance-Based Codes and Fire Safety Design Methods,* International Conference of Building Officials, Whittier, CA, 1998.

Computer Fire Models

Custer, R. L. P., & Meacham, B. J. "Introduction to Performance-Based Fire Safety," National Fire Protection Association, Quincy, MA, 1997.

Friedman, R. "An International Survey of Computer Models for Fire and Smoke," *Journal of Fire Protection Engineering,* vol. 4, no. 3, pp. 81–92, 1992.

Walton, W. D., & Budnick, E. K. "Deterministic Computer Fire Models," *Fire Protection Handbook,* 18th Ed., National Fire Protection Association, Quincy, MA, 1997.

Design Fire Scenarios

Alpert, R., & Ward, E. J. "Evaluation of Unsprinklered Fire Hazards," *Fire Safety Journal,* vol. 7, pp. 127–143, 1984.

Babrauskas, V. "Burning Rates," *The SFPE Handbook of Fire Protection Engineering,* 2nd Ed., National Fire Protection Association, Quincy, MA, 1995.

Babrauskas, V. "Table and Charts," *Fire Protection Handbook,* 18th Ed., National Fire Protection Association, Quincy, MA, 1997.

Babrauskas, V. "Will the Second Item Ignite?" *Fire Safety Journal,* vol. 4, pp. 281–292, 1981–1982.

Babrauskas, V., & Grayson, S. J. *Heat Release in Fires,* Elsevier Applied Science, New York, NY, 1992.

Cooper, L. Y. "Compartment Fire-Generated Environment and Smoke Filing," *The SFPE Handbook of Fire Protection Engineering,* 2nd Ed., National Fire Protection Association, Quincy, MA, 1995.

Custer, R. L. P. "Dynamics of Compartment Fire Growth," *Fire Protection Handbook,* 18th Ed., National Fire Protection Association, Quincy, MA, 1997.

Custer, R. L. P., "Introduction to the Use of Fire Dynamics in Performance-Based Design," *Proceedings of the Technical Symposium: Applications of Fire Dynamics,* The Society of Fire Protection Engineers and the University of British Columbia, Bethesda, MD, 1995.

Custer, R. L. P., & Meacham, B. J. *Introduction to Performance-Based Fire Safety,* National Fire Protection Association, Quincy, MA, 1997.

Drysdale, D. D. *An Introduction to Fire Dynamics,* 2nd Ed., John Wiley & Sons, Chichester, UK, 1999.

Drysdale, D. D. "Chemistry and Physics of Fire," *Fire Protection Handbook,* 17th Ed., National Fire Protection Association, Quincy, MA, 1991.

Gottuk, D. T., & Roby, R. J. "Effect of Combustion Conditions on Species Production," *The SFPE Handbook of Fire Protection Engineering,* 2nd Ed. National Fire Protection Association, Quincy, MA, 1995.

Hartzell, G. "Combustion Products and Their Effect on Life Safety," *Fire Protection Handbook,* 17th Ed., National Fire Protection Association, Quincy, MA, 1991.

Mulholland, G. W. "Smoke Production and Properties," *The SFPE Handbook of Fire Protection Engineering*, lst Ed., National Fire Protection Association, Quincy, MA, 1988.

NFPA 72, *National Fire Alarm Code*, National Fire Protection Association, Quincy, MA, 1999.

NFPA 204, *Guide for Smoke and Heat Venting*, National Fire Protection Association, Quincy, MA, 1998.

NFPA 921, *Guide for Fire and Explosion Investigation*, National Fire Protection Association, Quincy, MA, 1995.

"Performance-Based Fire Scenarios," *Primer #3*, National Fire Protection Association, Quincy, MA, 1998.

Quintiere, J. "Fundamentals of Enclosure Fire 'Zone' Models," *Journal of Fire Protection Engineering*, vol. 1, pp. 99–119, 1989.

Tewarson, A., "Generation of Heat and Chemical Compounds in Fires," *The SFPE Handbook of Fire Protection Engineering*, 2nd Ed., National Fire Protection Association, Quincy, MA, 1995.

Fire Barrier Damage and Structural Integrity

Fleishmann, C. "Analytical Methods for Determining Fire Resistance of Concrete Members," *The SFPE Handbook of Fire Protection Engineering*, National Fire Protection Association, Quincy, MA, 1995.

Milke, J. "Analytical Methods for Determining Fire Resistance of Steel Members," *The SFPE Handbook of Fire Protection Engineering*, National Fire Protection Association, Quincy, MA, 1995.

White, R. "Analytical Methods for Determining Fire Resistance of Timber Members," *The SFPE Handbook of Fire Protection Engineering*, National Fire Protection Association, Quincy, MA, 1995.

Fire Spread

Drysdale, D. D. *An Introduction to Fire Dynamics*, 2nd Ed. John Wiley & Sons, Chichester, UK, 1999.

Quintiere, J. Q. "Surface Flame Spread," *The SFPE Handbook of Fire Protection Engineering*, National Fire Protection Association, Quincy, MA, 1995.

Fire Testing

Babrauskas, V. "The Cone Calorimeter," *The SFPE Handbook of Fire Protection Engineering,* 2nd Ed., National Fire Protection Association, Quincy, MA, 1995.

Babrauskas, V., & Grayson, S. J. *Heat Release in Fires,* Elsevier Applied Science, New York, NY, 1992.

ISO 9705, *Full Scale Room Test,* International Standards Organization, Geneva, 1992.

Janssens, M., "Calorimetry," *The SFPE Handbook of Fire Protection Engineering,* 2nd Ed., National Fire Protection Association, Quincy, MA, 1995.

Goals and Objectives

"Goals, Objectives & Performance Criteria," *Primer #1,* National Fire Protection Association, Quincy, MA, 1998.

Human Behavior

Bryan, J. L. "Behavioral Response to Fire and Smoke," *The SFPE Handbook of Fire Protection Engineering,* National Fire Protection Association, Quincy, MA, 1995.

Ignition of Objects (Targets)

Kanury, A. M. "Flaming Ignition of Solid Fuels," *The SFPE Handbook of Fire Protection Engineering,* National Fire Protection Association, Quincy, MA, 1995.

Tewarson, A., "Generation of Heat and Chemical Compounds in Fires," *The SFPE Handbook of Fire Protection Engineering,* National Fire Protection Association, Quincy, MA, 1995.

Occupant Response

Proulx, G. "The Time Delay to Start Evacuation: Review of Five Case Studies," *Proceedings of the Fifth International Symposium on Fire Safety Science,* Y. Hasemi, Ed., International Association for Fire Safety Science, 1997.

Movement of People

Nelson, H. E., & MacLennan, H. A. "Emergency Movement," *The SFPE Handbook of Fire Protection Engineering,* National Fire Protection Association, Quincy, MA, 1995.

Pauls, J. "Movement of People," *The SFPE Handbook of Fire Protection Engineering,* National Fire Protection Association, Quincy, MA, 1995.

Smoke Damage

Tewarson, A. "Generation of Heat and Chemical Compounds in Fires," *The SFPE Handbook of Fire Protection Engineering,* National Fire Protection Association, Quincy, MA, 1995.

Thermal Effects

Engineering Guide to Assessing Flame Radiation to External Targets from Liquid Pool Fires, Society of Fire Protection Engineers, Bethesda, MD, 1999.

Engineering Guide to Predicting 1st and 2nd Degree Skin Burns, Society of Fire Protection Engineers, Bethesda, MD, 2000.

Purser, D. A. "Toxicity Assessment of Combustion Products," *The SFPE Handbook of Fire Protection Engineering,* National Fire Protection Association, Quincy, MA, 1995.

Toxicity

Purser, D. A. "Toxicity Assessment of Combustion Products," *The SFPE Handbook of Fire Protection Engineering,* National Fire Protection Association, Quincy, MA, 1995.

Verification Methods

"Verification Methods," *Primer #4,* National Fire Protection Association, Quincy, MA, 1998.

Visibility

Bryan, J. L. "Behavioral Response to Fire and Smoke," *The SFPE Handbook of Fire Protection Engineering,* National Fire Protection Association, Quincy, MA, 1995.

Drysdale, D. D. *An Introduction to Fire Dynamics,* John Wiley & Sons, 2nd Ed., Chichester, UK, 1999.

Mulholland, G. W. "Smoke Production and Properties," *The SFPE Handbook of Fire Protection Engineering,* National Fire Protection Association, Quincy, MA, 1995.

Purser, D. A. "Toxicity Assessment of Combustion Products," *The SFPE Handbook of Fire Protection Engineering,* National Fire Protection Association, Quincy, MA, 1995.

Heat Release Rate Data

Babrauskas, V. "Burning Rates," *The SFPE Handbook of Fire Protection Engineering,* 2nd Ed. National Fire Protection Association, Quincy, MA, 1995.

Babrauskas, V., Lawson, J. R., Walton, W. D., & Twilley, W. H. *Upholstered Furniture Heat Release Rates Measured with a Furniture Calorimeter,* NBSIR 82-2604, National Bureau of Standards, Gaithersburg, MD, 1982.

CIBSE Guide E, Fire Engineering, CRC Ltd., London, 1997.

Fire Engineering Guidelines, Fire Code Reform Centre Limited, Sydney, NSW, Australia, 1996.

"Fire Safety Engineering in Buildings," DD 240, British Standards Institute, 1997.

NFPA 92B, *Guide for Smoke Management Systems in Malls, Atria, and Large Areas,* National Fire Protection Association, Quincy, MA, 1995.

NFPA 130, *Standard for Fixed Guideway Transit Systems,* National Fire Protection Association, Quincy, MA, 1997.

"Standard Guide for Data for Fire Models," E1591, American Society for Testing and Materials, West Conshohocken, PA, 1999.

Appendix B

Example of Defining Objectives and Setting Performance Criteria

In this example, for compliance with existing regulations, a stakeholder must provide complete automatic sprinkler coverage or an alternate design that results in an *equivalent* level of safety. In order to develop a design that provides an *equivalent* level of safety, one must first understand the context of the requirement (i.e., establish boundary conditions), then quantify the safety level provided by the regulatory requirement (in a means acceptable to the stakeholders), develop an alternate design, quantify the safety level provided by the alternate design, and compare the safety level provided by the alternate design and the regulation-specified design.

To establish the context of the requirement, one can apply the approach of establishing goals (see Chapter 5).

In this example, assume that the requirement for complete automatic sprinkler protection primarily addresses life safety concerns and secondarily addresses property protection concerns.

The goal, therefore, might be to minimize fire-related injuries and to prevent undue loss of life. In meeting this goal, it might be assumed that the sprinkler cannot activate fast enough to prevent injury to, or even death of, person(s) in direct contact with the first materials burning. In this case, the objective might be to provide adequate time for those people outside of the room of fire origin to reach a place of safety without being overcome by the effects of fire and fire effluents. (For the purpose of this example, property protection goals and objectives will be ignored. However, in an actual analysis, property protection might have to be considered if it were considered by the regulation.)

The assumed life safety objective (i.e., provide adequate time for those people outside of the room of fire origin to reach a place of safety without being overcome by the effects of fire and fire effluents) can be accepted by the stakeholders as an adequate description of the safety level intended by the regulation, or the assumed objective can be tested by the application of probabilistic or deterministic engineering analysis to the code-prescribed parameters. For this example, the assumed life safety objective has been agreed to as an adequate description of the safety level intended by the regulation.

Further examples of fire protection goals and stakeholder objectives are provided in Table B–1.

Table B–1   Example of Fire Protection Goals and Related Stakeholder Objectives

Fire Protection Goals	Stakeholder Objectives
Minimize fire-related injuries and prevent undue loss of life	Provide adequate time for those people not intimate with the first materials burning to reach a place of safety without being overcome by the effects of fire and fire effluents
	Provide adequate time for those people outside the room or compartment of fire origin to reach a place of safety without being overcome by the effects of fire and fire effluents
	Provide adequate time for those people outside the floor of fire origin to reach a place of safety without being overcome by the effects of fire and fire effluents
Minimize fire-related damage to the building, its contents, and its historical features and attributes	Limit fire development and spread so that the structural integrity of the building is not threatened
	Limit fire development and spread to the compartment of fire origin
	Limit the spread of fire effluents to the floor of fire origin
	Limit fire spread to the building of fire origin
	Limit fire-related damage to a maximum of $500,000

Fire Protection Goals	Stakeholder Objectives
Minimize undue loss of operations and business-related revenue due to fire-related damage	Limit fire development and spread to the item of fire origin
	Limit the development and spread of fire so that the maximum fire-related process downtime is no greater than 24 hours
	Provide appropriate fire protection measures so that fires originating outside of operation-critical equipment do not cause damage to it
Limit the environmental impact of fire and fire protection measures	Prevent groundwater contamination by runoff of toxic materials as a consequence of the fire or of fire suppression activity
	Minimize air contamination that might result from the combustion of the building and its contents

At this point, the engineer and the stakeholders have agreed that the stakeholder objective is to provide adequate time for those people outside of the room of fire origin to reach a place of safety without being overcome by the effects of fire and fire effluents. However, they have not yet agreed on how this stakeholder objective might be achieved. To achieve this stakeholder objective, specific design objectives are required.

For this example, assume that if the room of fire origin does not flash over, then fire spread beyond the room of fire origin is minimized. Also, if the fire does not spread beyond the room of fire origin, then the conditions in the surrounding spaces are likely to remain tenable. Thus, if flashover is prevented, fire spread beyond the room of fire origin is most likely prevented, untenable conditions outside of the room of fire origin are unlikely, and the stakeholder objective will likely be met. Taking these assumptions to be valid, a design objective for this example might be to prevent flashover in the room of fire origin. Other possible design objectives might be as follows:

1. Minimize the likelihood of fire spread beyond the room of fire origin. In this case, flashover might occur within the room of fire origin so long as the fire is contained to that room.

2. Detect the fire early enough that occupants of the building can be alerted to the situation and reach a place of safety before the room flashes over.

3. Detect the fire early enough that the fire brigade (i.e., fire department or local brigade) can respond and take action to contain the fire to the room of fire origin.

In each of the above cases, a design objective was identified in terms that can be evaluated in a probabilistic or deterministic manner—flashover, resistance to fire spread, fire detection, egress time, and fire brigade response. This ability to describe design objectives in quantifiable terms is what differentiates design objectives from more qualitative stakeholder objectives.

Table B–2 contains additional examples of design objectives.

Once the design objectives are established, the performance criteria must be set. Consider the design objective to prevent flashover in the room of origin. Various references indicate that flashover can be characterized by an upper gas layer temperature of about 600 °C (1,100 °F)[1,2] and a heat flux at floor level of at least 20 kW/m^2 (1.8 Btu/ft^2-s). However, other references point out that although a range of 500–600 °C (930–1,100 °F) is most widely used, gas temperatures in the range of 300–650 °C (575–1,200 °F) have been associated with flashover.[3] This range in gas temperatures is due to a number of factors, such as compartment geometry and volume, fuel type, loading characteristics, and compartment ventilation. Likewise, different methods for estimating upper layer temperatures yield similar but different results.

In another example, maintaining egress path tenability might require both sprinklers and smoke management. In this case, the performance criteria might be sprinkler operation before the fire reaches 250 kW (240 Btu/s) and maintenance of a pressure gradient between the egress path and the areas containing smoke of 2.5 Pa (0.01 inches of water pressure) or higher. If the performance criteria included

Table B–2   Fire Protection Goals, Stakeholder Objectives, and Design Objectives

Fire Protection Goals	Stakeholder Objectives	Designs Objective
Minimize fire-related injuries and prevent undue loss of life	No loss of life outside of the room or compartment of fire origin	Prevent flashover in the room of fire origin
Minimize fire-related damage to the building, its contents, and its historical features and attributes	No significant thermal damage outside of the room or compartment of fire origin	Minimize the likelihood of fire spread beyond the room of fire origin
Minimize undue loss of operations and business-related revenue due to fire-related damage	No process downtime exceeding eight hours	Limit the smoke exposure to less than would result in unacceptable damage to the target
Limit environmental impacts of fire and fire protection measures	No groundwater contamination by fire suppression water runoff	Provide a suitable means for capturing fire protection water runoff

the application of the suppression agent before the heat release rate reached 80 watts (0.08 Btus/s) (about the heat release rate from a wooden match) in an office occupancy, the system might be prohibitively expensive. However, in a telecommunications facility, the damage due to corrosive combustion products might already be done by the time a fire reaches 80 watts. In this case, the added costs of materials control and early-warning air sampling detection might be needed.

Table B–3 contains examples of performance criteria.

Table B–4 suggests a framework for developing a matrix that identifies the various stakeholders and their goals and objectives. In some cases, assigning a weight or priority to the goals might be necessary. Some occupancies will place a higher priority on life safety as opposed to other goals (e.g., a hospital versus a warehouse).

Table B–3   Examples of Fire Protection Goals, Stakeholder Objectives, Design Objectives, and Performance Criteria

Fire Protection Goals	Stakeholder Objectives	Design Objectives	Performance Criteria
Minimize fire-related injuries and prevent undue loss of life	No loss of life outside of the room or compartment of fire origin	Prevent flashover in the room of fire origin	COHb level not to exceed 12 percent Visibility greater than seven meters (23 ft)
Minimize fire-related damage to the building, its contents, and its historical features and attributes	No significant thermal damage outside of the room or compartment of fire origin	Minimize the likelihood of fire spread beyond the room of fire origin	Upper layer temperature no greater than 200 °C (390 °F)
Minimize undue loss of operations and business-related revenue due to fire-related damage	No process downtime exceeding eight hours	Limit the smoke exposure to less than would result in unacceptable damage to the target	HCl no greater than five ppm Particulate no greater than 0.5 g/m^3
Limit environmental impacts of fire and fire protection measures	No groundwater contamination by fire suppression water runoff	Provide a suitable means for capturing fire protection water runoff	Impoundment capacity at least 1.20 times the design discharge

Table B–4   Matrix for Identifying Stakeholders and Their Goals and Objectives

Stakeholder	Fire Safety Goals (see Chapter 5)	Fire Safety Objectives (see Chapter 6)	Performance Criteria (see Chapter 7)
Building owner			
Building manager			
Design team			
Authorities having jurisdiction			
Fire			
Building			
Insurance			
Accreditation agencies			
Construction team			
Construction manager			
General contractor			
Subcontractors			
Tenants			
Building operations and maintenance			
Fire service			

References Cited

1. Drysdale, D. D. *An Introduction to Fire Dynamics*, 2nd Ed., John Wiley & Sons, Chichester, UK, 1999.

2. Walton, W., & Thomas, P. "Estimating Temperatures in Compartment Fires," *The SFPE Handbook of Fire Protection Engineering*, 2nd Ed., National Fire Protection Association, Quincy, MA, 1995.

3. *NFPA's Future in Performance-Based Codes and Standards*, National Fire Protection Association, Quincy, MA, 1995.

Appendix C

Use of Statistical Data to Choose Likely Fire Scenarios

Use of statistical data to choose likely fire scenarios can be challenging. Judgment is involved in selecting the categories that define alternative fire scenarios.

First, the dimensions and factors used to code and describe historical fires in available databases are not always the same dimensions and factors required as input data for a fire hazard analysis model. A translation is required, and every translation involves an assumed or demonstrated relationship between fire experience data characteristics and desired input data characteristics. These relationships must be substantiated as much as possible.

For example, it might not be valid to assume that all arson fires are fast or severe (i.e., have a steeply rising heat release rate or a high peak rate of heat release). Most arson fires do not involve the use of accelerants and are started by juveniles rather than individuals sophisticated in fire ignition. For fire scenario purposes in a fire hazard analysis, an arson fire might behave like an ordinary trash fire. By contrast, it might be valid to assume that all fires starting with the ignition of flammable liquids, whether accidental or deliberate, show the heat release rate curves associated with those products.

Second, when selecting likely fire scenarios, the analyst might identify fire scenarios that collectively account for a large share, preferably most, of the relevant fires. However, fire hazard analysis calculations require input data with so many characteristics defined that no one fire scenario can be expected to account for more than a tiny percentage of total fires. In this situation, the selection of fire scenarios should be done in two stages. The first stage might be partitioning all the fires that can occur into a manageable number of relatively homogeneous fire scenario classes. A fire scenario class might be a group of similar fire scenarios that are expected or known to yield similar severities when they occur. The second stage consists of selecting a representative fire within the class.

In a typical building, a manageable fire scenario class for analyzing home fires might consist of all fires originating in ordinary combustibles in any room that people normally occupy (i.e., excluding means of egress that people pass through but do not normally stay in and excluding service areas and concealed spaces). Then, the representative fire within the class might be an upholstered furniture fire within a living room, which is not necessarily the most likely fire scenario within the class (e.g., kitchen fires involving food on a stove are more common), but it is a common fire scenario that is easier to quantify for analysis purposes, given existing laboratory data, and is probably more typical of the fires in that class.

As the previous examples suggest, high-challenge fire scenarios can often be defined in terms of an area of origin and a fire size, both of which are dimensions that can be assessed statistically and used as engineering specifications for an engineering analysis of a building. Another appropriate dimension to use in defining high-challenge fire scenarios might be time of day, an indicator of the status of occupants.

Three major databases are available to analyze patterns in U.S. fire experience—the annual NFPA survey of fire departments, the FEMA/USFA National Fire Incident Reporting System (NFIRS), and the NFPA Fire Incident Data Organization (FIDO). Together, these three databases can provide information that may provide a basis for constructing fire scenarios.

Annual NFPA Survey of Fire Departments

The NFPA survey is based on a stratified random sample of roughly 3,000 U.S. fire departments (or roughly one of every ten fire departments in the country). The survey collects the following information that might be useful in fire hazard calculations: the total number of fire incidents, civilian deaths, civilian injuries, and the total estimated property damage (in dollars) for each of the major property-use classes defined by NFPA 901, *Standard for Fire Incident Reporting*.[1] These totals are analyzed and reported in NFPA's annual study, "Fire Loss in the United States," which traditionally appears in the September/October issue of *NFPA Journal*.

The NFPA survey is stratified by the size of population protected to reduce the uncertainty of the final estimate. Small, rural communities protect fewer people per department and are less likely to respond to the survey, so a larger number must be surveyed to obtain an adequate sample of those departments. NFPA also makes follow-up calls to a sample of the smaller fire departments that do not respond in order to confirm that those that responded are truly representative of smaller fire departments. On the other hand, large city departments are so few in number and protect such a large proportion of the population that it makes sense to survey all of them. Most respond, resulting in precision for their part of the final estimate.

These methods have been used in the NFPA survey since 1977. Because of the attention paid to representativeness and appropriate weighting formulas for projecting national estimates, the NFPA survey provides a basis for measuring national trends in fire incidents, civilian deaths, injuries, and direct property loss as well as for determining patterns and trends by community size and major region.

FEMA/USFA's National Fire Incident Reporting System (NFIRS)

The Federal Emergency Management Agency's U.S. Fire Administration (FEMA/USFA) distributes NFIRS, an annual computerized database of fire incidents with data classified according to a standard format based on NFPA 901. Roughly three-fourths of all states have NFIRS coordinators who receive fire incident data from participating fire departments and combine the data into a database. This data is then transmitted to FEMA/USFA. (To obtain a copy of the NFIRS database for a particular year, contact the National Fire Data Center, U.S. Fire Administration, 16825 South Seton Avenue, Emmitsburg, MD 21727-8995, or call 301-447-6771.) Participation by the states and the local fire departments within those states is voluntary. NFIRS captures roughly one-third to one-half of all U.S. fires each year. A larger proportion of U.S. fire departments are listed as participants in NFIRS, but not all departments provide data every year.

NFIRS provides the most detailed incident information of any reasonably representative national database not limited to large fires. NFIRS is the only database capable of addressing national patterns for fires of all sizes by specific property use and specific fire cause. The NFPA survey separates fewer than 20 of the hundreds of property-use categories defined by NFPA 901 and provides no cause-related information except for incendiary and suspicious fires. NFIRS also captures information on the construction type of the involved building, height of the building, extent of flame and smoke spread, performance of detectors and sprinklers, and victim characteristics, the latter in individual casualty reports that accompany the incident reports in an NFIRS file.

One weakness of NFIRS is that its voluntary character produces annual samples of shifting composition. Despite the fact that NFIRS draws on three times as many fire departments as the NFPA survey, the NFPA survey is more suitable as a basis for projecting national estimates because its sample is truly random and systematically stratified to be representative.

Most analysts use NFIRS to calculate percentages (e.g., the percentage of residential fires that occur in apartments or the percentage of apartment fire deaths that involve discarded cigarettes), which are then combined with NFPA-survey-based totals to produce estimates of fires, deaths, injuries, and dollar loss for subparts of the fire problem. This is the simplest approach now available to compensate for NFIRS' main weakness. It has been documented as an analysis method in a 1989 article in *Fire Technology.*[2]

NFPA's Fire Incident Data Organization (FIDO)

NFPA's FIDO is a computerized index and database that provides the most detailed incident information available, short of a full-scale fire investigation. The fires covered are those NFPA deems of major technical interest. The tracking system that identifies fires for inclusion in FIDO is believed to provide virtually complete coverage of incidents reported to fire departments involving three or more civilian deaths, one or more fire fighter deaths, or large dollar loss. It is redefined periodically to reflect the effects of inflation and has been defined since the late 1980s as $5,000,000 or more in direct property damage.

FIDO covers fires from 1971 to date, contains information on more than 70,000 fires, and adds about 2,000 fires per year. NFPA becomes aware of possible candidates for FIDO through a newspaper-clipping service, insurers' reports, state fire marshals, NFIRS, responses to the NFPA annual survey, and other sources. Once notified of a candidate fire, NFPA seeks standardized incident information from the responsible fire department and solicits copies of other reports prepared by concerned parties, such as the fire department's own incident report and the results of any investigations.

The strength of FIDO is its depth of detail on individual incidents. Coded information that might be captured by FIDO, but never by NFIRS, includes detailed types and performances of built-in systems for detection, suppression, and control of smoke and flame; detailed factors contributing to flame and smoke spread; estimates of time between major events in fire development (e.g., ignition to detection, detection to alarm); reasons for unusual delays at various points; indirect loss and detailed breakdowns of direct loss; and escapes, rescues, and numbers of occupants. Additional uncoded information often is available in the hard-copy files, which are indexed by FIDO for use in research and analysis.

FIDO is a resource that might be used with NFIRS-based national estimates in the same manner that NFIRS is used with the NFPA survey to produce national estimates. That is, a FIDO analysis might provide reasonable estimates of how a block of fires, estimated through NFIRS-based national estimates, further subdivides into more detailed categories.

Neither FIDO nor any other special fire incident database with detail exceeding that in NFIRS is directly available to analysts outside the organization that maintains the database. It will be necessary to work with analysts in charge of the database to obtain needed analyses, and at that time, the database analysts can help indicate what analyses are possible with the data.

Working with the Strengths and Weaknesses of Different Databases

No fire database can possibly capture all instances of unwanted fires. Few databases cover fires that are not reported to fire departments. By their nature, fire databases are biased in favor of *failures* rather than *successes*. The fire that is controlled so quickly it does not need to be reported to a fire department is not captured by the databases that cover reported fires. Analyses of the impact of devices and procedures that provide early detection or suppression also need to allow for the phenomena of missing *success* stories.

Quality control for a database is also an issue. For databases with limited depth of detail, like the NFPA survey, or limited breadth of coverage, like FIDO, which is primarily devoted to large fires, it is possible to invest considerable effort in ensuring that each report is as complete and accurate as possible. Follow-up calls can be used to fill gaps and check possible odd answers. For a database with the depth and breadth of NFIRS, however, the same level of quality control effort has not been possible. Consequently, NFIRS is missing many entries and has more that are dubious. The trade-off between data quality and data quantity is never easy; an analyst needs to be aware of the strengths and limitations of the sources before conducting an analysis.

References Cited

1. NFPA 901, *Standard Classifications for Incident Reporting and For Incident Data*, National Fire Protection Association, Quincy, MA, 1995.

2. Hall, J. R., Jr., & Harwood, B. "The National Estimates Approach to U.S. Fire Statistics," *Fire Technology,* vol. 25, no. 2, pp. 99–113, 1989.

Examples of Fire Scenario and Design Fire Scenario Identification

The examples below are intended to illustrate concepts presented in Chapter 8.

Fire Scenario

In a wastepaper basket, a fire starts due to a discarded cigarette, which smolders for several minutes and eventually leads to ignition of the contents and wastebasket in a hospital waiting room. This fire radiates sufficient energy to ignite an adjacent couch constructed of wood and polyurethane foam. The fire grows enough to ignite an adjacent wooden wardrobe that provides sufficient heat to cause the room to flashover. Most nonstaff occupants are nonambulatory and are not familiar with the exits or evacuation procedures. The staff has been trained with regard to emergency procedures during a fire incident.

Tools
Failure Analysis

Consider the design of a smoke detection system and a failure analysis to identify ways in which the detection system might not perform as expected. Ambient conditions, such as thermal layering, local heat sources, or the air flow generated by the HVAC system, might prevent smoke from reaching the detector. Thermal layering (e.g., in a tall space such as an atrium) might be sufficient to prevent an automatic sprinkler head from operating with a fire at the first-floor level. Maintenance conditions might also affect the ability of a detection system to perform as expected.

Each of these possibilities might then be considered as parts of fire scenarios. If these fire scenarios were used as design fire scenarios, the impact of these possible failures might be considered. If the detection system were used as part of a larger system such as one designed for smoke management or the discharge of special suppression agents, the impact of these failures might need to be considered. If these failures resulted in a failure to achieve design objectives, these scenarios would identify the need for alternate detection strategies of failure prevention or mitigation techniques.

"What If?" Analysis

In another example, consider the following question: "What happens if door A is open during a fire?" The following answers might apply: "Nothing, because the door connects to an isolated space not connected to the rest of the building," "One exit will be blocked," or "The fire will grow dramatically due to ignition of a more flammable fuel array in the adjacent space."

Fire Frequencies

In some applications, the fire initiation frequency might be based on the building floor area. For simplicity in this example, a homogeneous fire frequency has been assumed. Actual fire frequencies might vary in different areas due to occupancy, use, utilities, time of day, and other factors. Table D–1 shows an example of determining the expected fire frequency for a three-story building from a given fire frequency stated on a per floor area basis. In this example, a fire frequency of $3.0/10,000$ fires/m^2-yr ($0.3/10,000$ fires/ft^2-yr) is used.

Table D–1   Fire Initiation Frequency for a Three-Story Building

	Floor Area (Meters)	Fire Initiation Frequency (Fires/Yr)
First floor	1,000	0.30
Second floor	1,000	0.30
Third floor	800	0.24
Total	**2,800**	**0.84**

For this example, the fire initiation frequency would be 0.84 fires per year. When developing and interpreting fire frequencies, don't assume any limitations or conditions other than those explicitly stated. In the example above, the value 0.84 fires per year is the estimated frequency of a fire of any size. If the value of interest were the frequency of a fire exceeding a defined size or loss threshold, such as the frequency of fires involving at least $100,000 losses or of fires with flame spread beyond the room of origin, then that must be explicitly stated, and the frequency will be lower.

A conditional probability is the likelihood of an event occurring given the occurrence of a prior event. For example, if a fire occurs, what is the conditional probability that flame spreads beyond the room of origin? For this example, assume it has been determined by separate analysis that the conditional probability of fire propagation outside of the room of fire origin, given the magnitude and location of the fire and the construction and fire protection features of the building, is approximately 0.1. Then, the frequency of fires with flame spread beyond the room of origin will be equal to the frequency of fires (of any size) times the conditional probability that a fire will have flame spread beyond the room of origin in the example; this would be $0.84 \times 0.1 = 0.084$.

In this example, the expectation is that a fire involving two or more rooms might occur every 11.9 years ($1/0.084 \text{ yr}^{-1}$). In practice, the two-room fire might occur after this period, or it could occur next week.

Reliability and Availability

When reported separately, the method for calculating the overall system failure probability might be as follows:

$$P_{\text{fail}} = 1 - P_A P_R$$

where P_A = probability that the system is available to operate

P_R = probability that if available, the system functions as designed

If a given system is specified to have an availability of 1.0 and a reliability of 0.9, then the system is expected to fail 10 percent of the time.

$$P_{\text{fail}} = [1 - (1.0)(0.9)] \cdot 100\% = 0.1 \cdot 100\% = 10\%$$

If the availability is 0.95 and the reliability is 0.9, the system failure probability is 14.5 percent.

$$P_{\text{fail}} = [1 - (0.95)(0.9)] \cdot 100\% = 0.145 \cdot 100\% = 14.5\%$$

Risk

Assume for a given facility, a $500,000 loss is expected due to a given fire occurring once every 1,000 years. For this facility, a monetary expression of the risk could be estimated as follows:

$$R_{\$} = C \cdot F = (\$500,000 \text{ per fire})\left(\frac{1 \text{ fire}}{1,000 \text{ years}}\right) = 500 \frac{\$}{\text{year}}$$

If a death were to occur in one out of four fires, then the risk to people could be estimated as follows:

$$R_d = C \cdot F = \left(\frac{1 \text{ death}}{4 \text{ fires}}\right)\left(\frac{1 \text{ fire}}{1,000 \text{ years}}\right) = 2.5 \times 10^{-4} \frac{\text{deaths}}{\text{year}}$$

If a death were assumed equivalent to a $1,000,000 dollar loss, a monetary expression of the total risk for the above example could be estimated as follows:

$$R_t = R_{\$} + R_d K_d = 500 \frac{\$}{\text{year}} + \left(2.5 \times 10^{-4} \frac{\text{deaths}}{\text{year}}\right)\left(\frac{1 \times 10^6}{\text{death}}\right) = 750 \frac{\$}{\text{year}}$$

This example illustrates the concept of risk in a simplified manner. In practice, there will be a range of possible consequences and associated frequencies as well as social and value decisions that require the input of all stakeholders. In some cases, the social and value decisions might be codified or influenced by persons outside of the initial stakeholder group. This complexity is addressed in Chapter 10.

Implied Risk

Preventing flashover is a fire protection method that has been used to limit the spread of fire and might be accomplished by limiting the heat release rate of objects. Assume that for a given room, flashover will not occur if the HRR is below a threshold value. If the design fire scenario is based on a single item with a HRR slightly below this threshold value, then a second item adjacent to the first will cause the flashover prevention strategy to fail. Implicit in the use of a single item as the design fire scenario is the unstated assumption that the likelihood of the two items being put together when a fire occurs is too low to be considered for design purposes.

Appendix E

Risk Analysis

Risk can be quantified on the following basis[1]:

$$\text{Risk} = \sum \text{Risk}_i = \sum (\text{Loss}_i \cdot F_i)$$

where Risk_i = Risk associated with scenario i

Loss_i = Loss associated with scenario i

F_i = Frequency of scenario i occurring

Also, reliability of protection measures (e.g., trial designs) can be included on the following basis:

$$\text{Risk} = \sum \text{Risk}_i = \sum \left[F_i \cdot \left(\text{Loss}_i \cdot R_k + \overline{\text{Loss}_i}(1 - R_k) \right) \right]$$

where $\overline{\text{Loss}_i}$ = Loss associated with scenario i if the trial design fails

R_k = Reliability of trial design

In addition, risk can be expressed as the frequency that a loss will exceed a given threshold, (e.g., a performance criterion). This would take the following form:

$$\text{Risk} = \sum \text{Risk}_i = \sum [F_i(\text{Loss}_i > n)]$$

where $F_i(\text{Loss}_i > n)$ is the frequency that a loss will exceed the threshold n (i.e., the set n of performance criteria).

In classical risk analysis, the overall risk is often obtained from the sum of the risks associated with individual potential scenarios of a specific type (e.g., fire or explosion). This can be illustrated using a simple three-room example. Figure E–1 shows an event tree for a three-room building that is subdivided as shown. For this example, the fire initiation frequency, F_i, is assumed to be uniformly distributed across the three rooms, and the consequence of a single-room loss is $C/3$. (The consequence of a fire involving all three rooms would be C.) If the probability that the fire will be contained in one room is P_c and the probability that it is prevented from propagating to the third is P_f, then the overall risk as shown in Figure E–1 is as follows:

$$R = \frac{C}{3}\left[F_I P_1 P_c\right] + \frac{2C}{3}\left[F_I P_1 (1 - P_c) P_f\right] + C\left[F_I P_1 (1 - P_c)(1 - P_f)\right]$$

$$+ \frac{C}{3}\left[F_I P_2 P_c\right] + \frac{2C}{3}\left[F_I P_2 (1 - P_c) P_f\right] + C\left[F_I P_2 (1 - P_c)(1 - P_f)\right]$$

$$+ \frac{C}{3}\left[F_I P_3 P_c\right] + \frac{2C}{3}\left[F_I P_3 (1 - P_c) P_f\right] + C\left[F_I P_3 (1 - P_c)(1 - P_f)\right]$$

where P_1, P_2, and P_3 are the probabilities that a fire will start in Room 1, 2, or 3, respectively. This equation simplifies as follows:

$$R = \frac{CF_i}{3}[3 - 2P_c - P_f + P_c P_f]$$

For this example, P_c and P_f can be interpreted as the success probabilities of the fire barriers. To place these results in context, numeric values will be added. If P_c and P_f are both equal to 0.1 (i.e., fire propagates nine in ten times), then the risk is as follows:

$$R = \frac{CF_i}{3}[3 - 2(0.1) - (0.1) + (0.1)(0.1)] = 0.90CF_i$$

If P_c and P_f are both equal to 0.9 (i.e., fire propagates one in ten times), then the risk is as follows:

$$R = \frac{CF_i}{3}[3 - 2(0.9) - (0.9) + (0.9)(0.9)] = 0.37CF_i$$

If P_c and P_f are set to unity (i.e., fire barriers never fail), then the risk is as follows:

$$R = \frac{CF_i}{3}[3 - 2(1) - (1) + (1)(1)] = \frac{CF_i}{3} \approx 0.33CF_i$$

Although this example is simplified, it does suggest how complicated a classical risk analysis can be. For each protective feature, the number of branches (i.e., potential outcomes) in the event tree will increase. This increase is usually geometric.

The example above also illustrates an important concept in risk-based calculations. The bounding risk for this problem would be CF_i (i.e., complete facility loss). This risk applies if all protective features are assumed to always fail. The risk when the protective features are always assumed to work (i.e., the fire barriers never fail; P_c and P_f are set to unity) is the lower bound risk. The potential range for the actual risk is thus bounded between $0.33CF_i$ and CF_i. The better the protection, the closer the risk will approach $0.33CF_i$.

The example above can also be used to illustrate the difference between fire scenarios and design fire scenarios. The fire scenarios for this example could consider the sequence of fire propagation from room to room (e.g., starting in Room 1, then going to Room 2, and finally to Room 3). There are 15 possible paths for the example above.

1. Starts in Room 1 and is contained in Room 1
2. Starts in Room 1 and propagates to Room 2, but not to Room 3
3. Starts in Room 1 and propagates to Room 3, but not to Room 2
4. Starts in Room 1 and propagates to Room 2 and then to Room 3
5. Starts in Room 1 and propagates to Room 3 and then to Room 2
6. Starts in Room 2 and is contained in Room 2
7. Starts in Room 2 and propagates to Room 1, but not to Room 3
8. Starts in Room 2 and propagates to Room 3, but not to Room 1
9. Starts in Room 2 and propagates to Room 1 and then to Room 3
10. Starts in Room 2 and propagates to Room 3 and then to Room 1
11. Starts in Room 3 and is contained in Room 3
12. Starts in Room 3 and propagates to Room 1, but not to Room 2
13. Starts in Room 3 and propagates to Room 2, but not to Room 1
14. Starts in Room 3 and propagates to Room 1 and then to Room 2
15. Starts in Room 3 and propagates to Room 2 and then to Room 1

If simultaneous propagation to the second and third room were considered a significant threat, there would be 18 scenarios. This scenario has not considered details such as doors, people, and fire timing.

This example demonstrates the level of effort that would be expended if each fire scenario were analyzed. There could be three design fire scenarios for this example—1-, 2-, and 3-room involvement. For this example (as it was presented), the actual path (i.e., sequence of involvement and timing) to this loss is unimportant and thus not part of the design fire scenario. If fire department intervention were credited, then a comparison between response time and fire propagation might need to be included. The intent is to ensure that the design fire scenarios encompass all credible scenarios. Those scenarios that are not encompassed by the selected design fire scenarios are considered to be an acceptable risk.

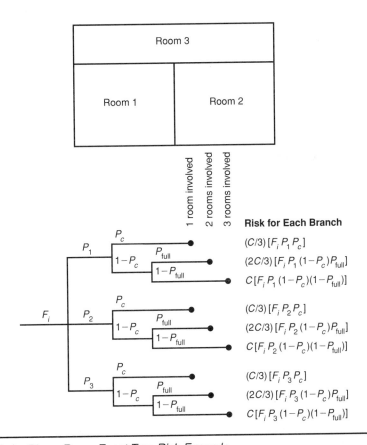

FIGURE E–1 *Three-Room Event Tree Risk Example*

Reference Cited

1. Custer, R. L. P., & Meacham, B. J. *Introduction to Performance-Based Fire Safety,* National Fire Protection Association, Quincy, MA, 1997.

Appendix **F**

Selecting Models or Other Analytical Methods*

A wide range of analytical tools are available to the engineer.[1,2] When selecting a particular calculation method, its predictive capability must be considered. Guidance for evaluating the predictive capability of computer fire models can be found in ASTM E1355-96.[3] The process of model evaluation is important for determining the appropriate use as well as the limitations for a particular application. Although a model might be appropriate for one fire scenario, it might be inappropriate for another.

Determining suitability for use also requires an evaluation of the sensitivity of the given model. Sensitivity analysis determines the effect of changes in individual input parameters on the results of a given model. The analysis might be done by holding all but one input variable constant and systematically studying the effects of that one variable on the predicted result.

If a high degree of uncertainty about the magnitude or variability of a given input variable for the design fire or the fire scenario exists, the model might be inappropriate if the output is particularly sensitive to that variable. On the other hand, if the given input has little effect on the output, then the tolerance for uncertainty

*Adapted from Custer, R. L. P., & Meacham, B. J. *Introduction to Performance-Based Fire Safety,* National Fire Protection Association, Quincy, MA, 1997.

is increased. In most cases, the most significant variable will be the heat release rate history input for the design fire. In some situations, however, heat losses to the compartment boundaries, specification of the physical properties of compartment boundaries, or the size of vent openings might significantly affect the result.

A model's ability to accurately predict outcomes can be assessed by comparison with standard tests or large-scale compartment fire tests. In addition, documented fire experience from actual eyewitness accounts or behavior of materials in actual fires can also be used. Another evaluation method might be comparison of model results with previously published data on full-scale tests in which the specific output parameters being evaluated have been measured. Outcomes such as structural failure, increase of temperature and smoke, available escape time, or the response of detection systems might be used as benchmarks for comparing models to test results.

The process of model selection should also include a review of any limitations placed on the use of the model, either in the applications manual or the supporting technical literature.

General Guidelines for Modeling Analysis[4]

The procedures for conducting a modeling analysis for a performance-based design will vary considerably, depending on the complexity of the analysis at hand. If the intent is to determine the size of the fire at the time of sprinkler operation or the length of time required for that operation, for example, the analysis could be performed using the sprinkler–detector subroutine in FPETool. On the other hand, if the objective is to determine the spacing of smoke detectors for a given design fire scenario that will allow a ten-minute, safe egress time from the building in the absence of automatic sprinklers, the analysis becomes more complex.

One way to organize complex modeling tasks is to prepare an outline or plan indicating the modeling steps that need to be accomplished to arrive at an evaluation of a trial fire protection design for a given design fire scenario. Once the model(s) of choice have been selected and documented, a modeling matrix can be developed. The modeling matrix is based on a list of the fire scenarios and design fires to be modeled and a group of trial fire protection designs to be evaluated. The variations included in the matrix might include presence or absence of specific fire protection systems and the location of damage targets, such as occupants or property. Each damage target will be characterized by a performance criterion, such as a minimum allowable exposure to heat, smoke, or corrosive agents. In addition, there might be multiple ventilation conditions, such as open and closed doors or on/off HVAC systems.

A full evaluation of a large modeling matrix for a particular design problem might take many trials. To facilitate record keeping, a naming convention should be developed for the data files. The modeling process should be treated as a laboratory experiment in which variables are altered one at a time and a lab notebook is kept.

The results should be reviewed as work progresses. Trends might be noted that can help focus the analysis. For example, if the performance criteria (untenable conditions) are reached prior to occupant escape time for small fires with a given detector spacing, it might be unnecessary to run larger fires unless a design involving reduced detector spacing or increased sensitivity is being evaluated.

Although each design situation is different, the review of published work for examples of how other engineers have used analysis and modeling tools is often useful. One place to look might be in the documentation for specific computer models.[5,6] Frequently, sample runs are provided to assist users who are trying to become familiar with the software. Studying the manner in which these runs were constructed can be a useful exercise. In addition, various articles have been written describing applications of modeling techniques to a variety of performance-based design problems.

Another source of sample applications of computer modeling is fire reconstruction and failure analysis.[7,8,9] A number of studies of actual fires have employed methodologies that might be useful for understanding computer applications and might be applied to performance-based design.

Limitations of Modeling

A number of limitations on the use of correlations and models for the prediction of fire phenomena exist.[1] The following is a brief overview of some of the restrictions or assumptions that limit the use of models. This overview is not exhaustive, so the reader is urged to review the technical documentation and references for models or correlations being used to determine what limitations might be present.

Room Geometry

Most fire models that deal with the prediction of ceiling layer thickness, ceiling jet velocity, or the operation of detection devices are based on the assumption of horizontal smooth ceilings. Thus, calculations with these programs might not model the effect, for example, of beams on the operation of sprinklers. The relative dimensions of compartments also are subject to some restrictions. For example, some models might not be appropriate for rooms with length-to-width ratios greater than 10:1 during early stages of fire growth or for compartments in which the height to minimum horizontal distance ratio exceeds one. As such, users are urged to exercise caution when evaluating fires in rooms larger than sizes that have been verified experimentally.[10] In addition, compartment layouts frequently have vent openings (e.g., doors and windows) on different wall surfaces. However, computer fire models generally treat the ventilation openings in a compartment as if they were one single opening, and these models do not address any issues that might arise due to the specific location of the given opening.

Interior Finishes

In general, most compartment fire models consider the thermal characteristics of the bounding surfaces of the compartment for the purposes of energy balance calculations. At the present level of model sophistication, the combustible nature or fuel contribution and flame spread effects of interior finish materials (e.g., walls, ceilings, and floors) are not included in the fire growth calculations unless they can be made a part of the overall heat release rate curve input by the modeler. However, some published works provide methods for calculating flame spread on wall-lining materials and for the resulting heat release rate in a compartment.[11,12,13]

Fire Suppression

Although the capability for predicting the effects of fire suppression activities or systems is not fully modeled by any of the programs currently in use, some programs, such as the fire simulator routine in FPETool and the FASTLite, do provide capabilities for the evaluation of sprinkler systems. However, these programs model the effect of only a single sprinkler operation on the heat release rate of the input design fire curve. Cooling effects on the hot gas layer and the effects of entrainment into the water spray are not included. Neither is the effect of prewetting material not yet ignited, which is lying within the spray envelope. Care should be taken in applying these sprinkler models when field experience might indicate that, due to the compartment geometry or the nature and geometry of the fuels involved, multiple sprinkler operation is expected.

The effects of nonwater-based suppression systems (e.g., water mist, carbon dioxide, or other gaseous agents) are not modeled. To assess the effectiveness of these suppression strategies for performance-based design, the engineer should review the literature with respect to the performance of the candidate agent in actual fire suppression tests that are similar to the design situation being evaluated. In some instances, arranging for testing to be conducted to obtain information to adequately model the performance of a system might be valuable.

Accuracy of Fire Models

The accuracy of a fire model might be assessed by its ability to predict the results of actual experimental data. Assessing models to determine their predictive capability is part of the process described earlier.

A number of published papers compare various fire models and experimental fire data.[14,15,16,17,18,19,20] When reviewing these papers, understanding the implications of the stated outcomes to the specific project under evaluation is important.

For example, Deal and Beyler[18] compared, measured, and predicted temperature rises using a variety of correlations. Their work indicated that some correlations overpredict temperature rise while others tend to underpredict. Understanding how variation between predicted and measured values affects the use of correlations and models for performance-based design or evaluation is important. If one were to select a correlation that overpredicted temperature, it might be said that this would be conservative and, in effect, provide a safety factor. Stated differently, using this correlation might predict higher than actual temperatures at a given time. This could be interpreted to mean that in actual situations, structural components would be subjected to less than predicted temperatures, as would unignited combustible materials or perhaps building occupants. If the correlation, however, is used to determine or predict when a sprinkler or detector might operate, high temperatures might not be conservative; the prediction might have the fire protection system operating sooner than it would in the actual situation. On the other hand, underpredicting temperature, for example, might result in higher temperatures in any point in time, resulting in greater thermal stresses on structural elements, materials, or occupants within the exposed area.

Nelson and Deal reported on an approach for appraising expected performance of fire models by comparison with actual compartment fire data.[19] In their demonstration of this methodology, Nelson and Deal found that the four models tested provided what they felt were reasonable approximations for the tests being evaluated. Temperature of the upper layer, oxygen concentration, interface height, and flow of products from the room out of a vent were evaluated. The results, however, did indicate that some models tended to underpredict while others overpredicted actual experimental data.

Other Limitations

Models have other limitations, such as prediction of a uniform temperature throughout the hot layer and the assumption that changes in temperature occur instantaneously throughout the volume of the layer. In addition to the model's documentation, other literature such as *The SFPE Handbook of Fire Protection Engineering* should be consulted to identify these limitations.

References Cited

1. Friedman, R. "An International Survey of Computer Models for Fire and Smoke," *Journal of Fire Protection Engineering,* vol. 4, no. 3, pp. 81–92, 1992.

2. *Catalog of Computer Fire Models,* Society of Fire Protection Engineers, Boston, MA, 1995.

3. ASTM E1355-96, *Standard Guide for Evaluating Predictive Capability of Fire Models,* American Society for Testing and Materials, Philadelphia, PA, 1996.

4. Custer, R. L. P., & Meacham, B. J. *Introduction to Performance-Based Fire Safety,* National Fire Protection Association, Quincy, MA, 1997.

5. Peacock, R. D., Jones, W. W., Bukowski, R. W., & Forney, C. L. *Technical Reference Guide for the HAZARD I Fire Hazard Assessment Method, Version 1.1,* National Institute of Standards and Technology, Gaithersburg, MD, 1991.

6. Portier, R. W., Peacock, R. D., & Reneke, P. A. *FastLite: Engineering Tools for Estimating Fire Growth and Smoke Transport,* Special Publication 899, National Institute of Standards and Technology, Gaithersburg, MD, 1996.

7. Nelson, H. E. *An Engineering Analysis of the Early Stages of Fire Development—The Fire at the DuPont Plaza Hotel and Casino—December 31, 1986,* NBSIR87-3560, National Institute of Standards and Technology, Gaithersburg, MD, 1987.

8. Nelson, H. E. "Engineering Analysis of Fire Development in the Hospice of Southern Michigan," *Proceedings of the International Association for Fire Safety Science,* Hemisphere Publishing Corporation, New York, NY, 1989.

9. Nelson, H. E. *Engineering View of the Fire of May 4, 1988 in the First Interstate Bank Building, Los Angeles, California,* NISTIR89-4061, National Institute of Standards and Technology, Gaithersburg, MD, 1985.

10. Portier, R. W., Reneke, P. E., Jones, W. W., & Peacock, R. D. *A User's Guide for CFAST, Version 1.6,* National Institute of Standards and Technology, Gaithersburg, MD, 1992.

11. Karlsson, B. "Models for Calculating Flame Spread on Wall Lining Materials and the Resulting Heat Release Rate in a Room," *Fire Safety Journal,* vol. 23, no. 4, pp. 365–386, 1994.

12. Mittler, H. E., & Stekler, K. D. *A Model of Flame Spread on Vertical Surfaces,* National Institute of Standards and Technology, Gaithersburg, MD, NISTIR 5619, 1995.

13. Dembsey, N. A., & Williamson, R. B. "Coupling the Fire Behavior of Contents and Interior Finishes for Performance-Based Fire Codes Evaluation of a Fire Spread Model," *Journal of Fire Protection Engineering,* vol. 8, no. 3, pp. 119–132, 1997.

14. Dembsey, N. A., Pagni, P. J., & Williamson, R. B. "Compartment Fire Experimental Data: Comparison to Models," *Proceedings of the International Conference on Fire Research and Engineering,* Society of Fire Protection Engineers, Boston, MA, 1995.

15. Wong, Doung, D. Q. "Accuracy of Computer Fire Models: Some Comparisons with Experimental Data with Australia," *Fire Safety Journal,* vol. 16, no. 6, pp. 415–431, 1990.

16. Peacock, R. F., Davis, S., & Babrauskas, V. "Data for Room Fire Model Comparisons," *Journal of Research of the National Institute of Standards and Technology,* vol. 96, no. 4, pp. 411–462, 1991.

17. Peacock, R. D., Jones, W. W., & Bukowski, R. W. "Verification of a Model of Fire and Smoke Transport," *Fire Safety Journal,* vol. 21, no. 2, pp. 89–129, 1993.

18. Deal, S., & Beyler, C. L. "Correlating Pre-Flashover Room Fire Temperatures," *Journal of Fire Protection Engineering,* vol. 2, no. 2, pp. 33–48, 1990.

19. Nelson, H. E., & Deal, S. "Comparing Compartment Fires with Compartment Fire Models," *Proceedings of the 3rd International Symposium of the International Association for Fire Safety Science,* Elsevier Applied Science, New York, 1991.

20. *DETACT Evaluation Report,* Society of Fire Protection Engineers, Bethesda, MD, 1998.

Appendix **G**

Uncertainty Analysis

For performance-based designs, an understanding of the uncertainties can be important. This appendix provides an overview of the methods that can be used to evaluate and combine uncertainties. Usually, the uncertainty will be dominated by a few key parameters. Significant contributors should be evaluated. A sensitivity analysis as discussed in Section 10.5.6.2 might identify these parameters that most significantly contribute to the uncertainty.

Steps in an Uncertainty Analysis

In order to determine the best way to treat the uncertainty, the following steps are taken:

- The most significant parameters should be identified.
- The types of uncertainty (e.g., theory or model) that exist should be identified.
- It should be determined if it is crucial to treat the uncertainty quantitatively. Only quantities about which the uncertainty has scientific significance (i.e., is capable of reversing the acceptability of the final solution) need to be treated quantitatively.
- The appropriate methodology or tool should be selected for the job.
- Uncertainty is encoded on crucial variables and propagated throughout the analysis.

Determining the Scientific Significance of an Uncertain Quantity

Treating every uncertainty quantitatively in a complex performance-based design might be impossible. The time required and computational complexity is prohibitive. However, it is unnecessary to do this. When dealing with uncertainty, one of the most important challenges is to identify and focus on those uncertainties that matter in understanding the project scope, and in making decisions about the fire safety tools, methods, and design options.

1. *Scientific Uncertainty versus Statistical Uncertainty. Statistically significant* refers to a mathematical calculation that verifies that two quantities are in fact the same or different. *Scientifically significant* refers to whether the difference is large enough to cause a change in the outcome criteria or final decision.

2. *Crucial variables* are defined as those whose uncertainties have the potential to change the acceptability of a solution. Several types of sensitivity analysis can be used to determine the scientific significance of an uncertainty and to identify the crucial variables. Some of these tools include importance analysis, parametric analysis, and switchover analysis as described in Section 10.5.6.

Selecting the Appropriate Approach or Tool for the Treatment of Uncertainty
An Approach to Scientific Uncertainty

Theory and Model Uncertainties. For an engineering correlation, uncertainties are identified by comparing and contrasting limitations and test conditions to the scenario being evaluated. For engineering calculations, assumptions of the model are compared to the fire scenario being modeled to determine any key differences. For correlations and calculations, comparisons of predictions to real-scale test results are often useful. When available, running various calculations with different models can help to determine the envelope within which the true answer lies.

Data and Model Inputs. They are often empirical parameters whose uncertainty can be treated probabilistically. Classical uncertainty analysis, as described in Section 10.5.4, handles this well.

Domain/Boundary Conditions, Level of Detail of Model. To determine how the boundary conditions of a model affect uncertainty, conduct multiple trials with a model under various boundary conditions. For example, model a large space whose boundaries are the entire building versus a model of a single draft-curtained area. To determine how the level of detail of the model affects uncertainty, perform calculations with more than one model. Run one or two cases with a more detailed model. If widely different results are obtained, the more detailed model should be used.

An Approach to Uncertainties in Human Behaviors, Risk Perceptions, Attitudes, and Values

First, one of the most important factors for dealing with these types of uncertainties is that all assumptions be made explicit in the analysis. Second, in these areas, one should be extra careful regarding *conservative* assumptions. For example, an assumed soot yield value might be conservative for smoke detector activation and not conservative, in fact the opposite, for life safety. Third, one should perform sensitivity analysis on all human behavioral assumptions, risk perceptions, attitudes, and values to determine assumptions critical to the design outcome. Fourth, how the results of a calculation change as a function of behavioral changes for these critical assumptions should be made explicit.

Tools for Classical Uncertainty Analysis

Random error is the result of variation or scatter in repeated measurements of a parameter. It is estimated using the standard deviation of a data set.

$$S = \sqrt{\sum_{k=1}^{N} \frac{(X_k - \bar{X})^2}{N - 1}}$$

where N is the number of measurements and $\bar{X}$ is the mean of the measurements. If there are multiple sets of measurements, the standard deviation of the sample mean can be used as the random estimate.

$$S_{\bar{X}} = \frac{S_X}{\sqrt{N}}$$

Systematic error, B, is constant for a repeated set of measurements. Based on engineering judgement, it is typically estimated at 95 percent confidence. If the systematic error is not reported, it is usually assumed to be equal to zero.

The total error, U, is the combination of the random error, S, and the systematic error, B. For 95 percent confidence, the equation is as follows:

$$U_{95} = 2 \sqrt{\left(\frac{B}{2}\right)^2 + \left(S_{\bar{X}}\right)^2}$$

The parameter will typically be reported as $\bar{X} \pm U$ (95 percent confidence).

If multiple parameters are combined to produce a resultant (i.e., a calculation), the uncertainty of the resultant can be estimated by combining the random and systematic uncertainties of each parameter used in the calculations as follows:

$$S_{\text{combined}} = \sqrt{\sum_{i=1}^{n} \theta_i S_i^2} \qquad B_{\text{combined}} = \sqrt{\sum_{i=1}^{n} \theta_i B_i^2}$$

where θ is a weighing factor derived from the equation used to produce the resultant.

Index

Note: Page numbers in **bold** type indicate definitions.